Value Proposition Design

Cover image: Pilot Interactive

Cover design: Alan Smith and Trish Papadakos

This book is printed on acid-free paper. ∞

Copyright © 2014 by Alexander Osterwalder, Yves Pigneur, Alan Smith, Greg Bernarda, and Patricia Papadakos. All rights reserved.

Published by John Wiley & Sons, Inc., Hoboken, New Jersey.

Published simultaneously in Canada.

No part of this publication may be reproduced, stored in a retrieval system, or transmitted in any form or by any means, electronic, mechanical, photocopying, recording, scanning, or otherwise, except as permitted under Section 107 or 108 of the 1976 United States Copyright Act, without either the prior written permission of the Publisher, or authorization through payment of the appropriate per-copy fee to the Copyright Clearance Center, 222 Rosewood Drive, Danvers, MA 01923, (978) 750-8400, fax (978) 646-8600, or on the web at www.copyright.com. Requests to the Publisher for permission should be addressed to the Permissions Department, John Wiley & Sons, Inc., 111 River Street, Hoboken, NJ 07030, (201) 748-6011, fax (201) 748-6008, or online at www.wiley.com/go/permissions.

Limit of Liability/Disclaimer of Warranty: While the publisher and author have used their best efforts in preparing this book, they make no representations or warranties with the respect to the accuracy or completeness of the contents of this book and specifically disclaim any implied warranties of merchantability or fitness for a particular purpose. No warranty may be created or extended by sales representatives or written sales materials. The advice and strategies contained herein may not be suitable for your situation. You should consult with a professional where appropriate. Neither the publisher nor the author shall be liable for damages arising herefrom.

For general information about our other products and services, please contact our Customer Care Department within the United States at (800) 762-2974, outside the United States at (317) 572-3993 or fax (317) 572-4002.

Wiley publishes in a variety of print and electronic formats and by print-on-demand. Some material included with standard print versions of this book may not be included in e-books or in print-on-demand. If this book refers to media such as a CD or DVD that is not included in the version you purchased, you may download this material at http://booksupport.wiley.com. For more information about Wiley products, visit www.wiley.com.

ISBN 978-1-118-96805-5 (cloth); ISBN 978-1-118-96807-9 (ebk); ISBN 978-1-118-96806-2 (ebk); ISBN 978-1-118-97310-3 (ebk)

Printed in the United States of America

10 9 8 7 6 5 4 3

How to create products and services customers want. Get started with...

Value Proposition Design

strategyzer.com/vpd

Written by
Alex Osterwalder
Yves Pigneur
Greg Bernarda
Alan Smith

Designed by
Trish Papadakos

WILEY

1. Canvas

1.1 Customer Profile *10*
1.2 Value Map *26*
1.3 Fit *40*

2. Design

2.1 Prototyping Possibilities *74*
2.2 Starting Points *86*
2.3 Understanding Customers *104*
2.4 Making Choices *120*
2.5 Finding the Right Business Model *142*
2.6 Designing in Established Organizations *158*

3. Test

3.1 What to Test *188*
3.2 Testing Step-by-Step *196*
3.3 Experiment Library *214*
3.4 Bringing It All Together *238*

4. Evolve

Create Alignment *260*
Measure & Monitor *262*
Improve Relentlessly *264*
Reinvent Yourself Constantly *266*
Taobao: Reinventing (E-)Commerce *268*

Glossary *276*
Core Team *278*
Prereaders *279*
Bios *280*
Index *282*

You'll love *Value Proposition Design* if you've been...

Overwhelmed by the task of true value creation

Sometimes you feel like...

- There should be better tools available to help you create value for your customers and your business.
- You might be pursuing the wrong tasks and you feel insecure about the next steps.
- It's difficult to learn what customers really want.
- The information and data you get from (potential) customers is overwhelming and you don't know how to best organize it.
- It's challenging to go beyond products and features toward a deep understanding of customer value creation.
- You lack the big picture of how all the puzzle pieces fit together.

Frustrated by unproductive meetings and misaligned teams

You have experienced teams that...

- Lacked a shared language and a shared understanding of customer value creation.
- Got bogged down by unproductive meetings with tons of unstructured "blah blah blah" conversations.
- Worked without clear processes and tools.
- Were focused mainly on technologies, products, and features rather than customers.
- Conducted meetings that drained energy and ended without a clear outcome.
- Were misaligned.

Involved in bold shiny projects that blew up

You have seen projects that...

- Were big bold bets that failed and wasted a lot money.
- Put energy into polishing and refining a business plan until it perpetuated the illusion that it could actually work.
- Spent a lot of time building detailed spreadsheets that were completely made up and turned out to be wrong.
- Spent more time developing and debating ideas rather than testing them with customers and stakeholders.
- Let opinions dominate over facts from the field.
- Lacked clear processes and tools to minimize risk.
- Used processes suited for running a business rather than ones for developing new ideas.

Disappointed by the failure of a good idea.

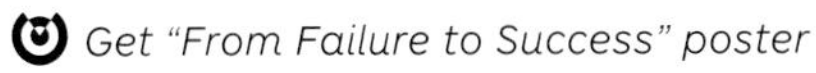

Get "From Failure to Success" poster

Value Proposition Design will help you successfully...

Understand the patterns of value creation

Organize information about what customers want in a simple way that makes the patterns of value creation easily visible. As a result, you will more effectively design value propositions and profitable business models that directly target your customers' most pressing and important jobs, pains, and gains.

Gain clarity.

Leverage the experience and skills of your team

Equip your team with a shared language to overcome "blah blah blah," conduct more strategic conversations, run creative exercises, and get aligned. This will lead to more enjoyable meetings that are full of energy and produce actionable outcomes beyond a focus on technology, products, and features toward creating value for your customers and your business.

Get your team aligned.

Avoid wasting time with ideas that won't work

Relentlessly test the most important hypotheses underlying your business ideas in order to reduce the risk of failure. This will allow you to pursue big bold ideas without having to break the bank. Your processes to shape new ideas will be fit for the task and complement your existing processes that help you run your business.

Minimize the risk of a flop.

Design, test, and deliver what customers want.

Get "From Failure to Success" poster

Our Value Proposition to You

The links you see on the side of every page point to resources in the online companion.

Watch for the Strategyzer logo and follow the link to online exercises, tools/templates, posters, and more.

Note: To gain access to these exclusive online portions of *Value Proposition Design*, you'll need to prove you own the book. Keep the book near you to help you answer the secret questions and verify your ownership!

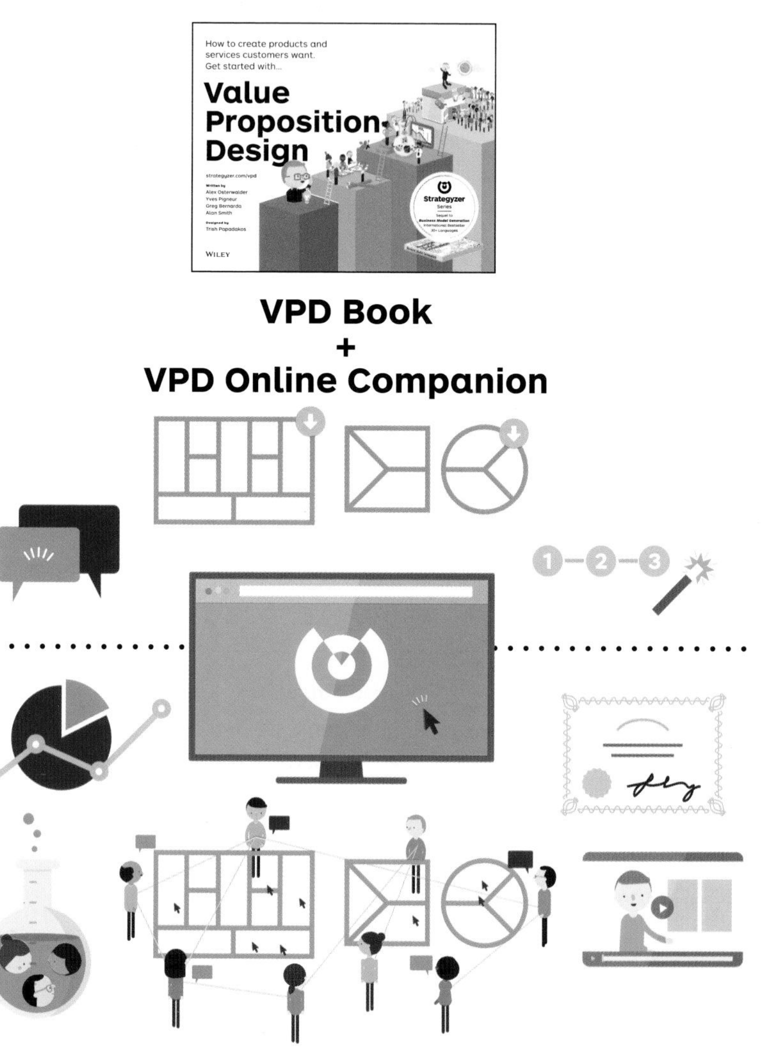

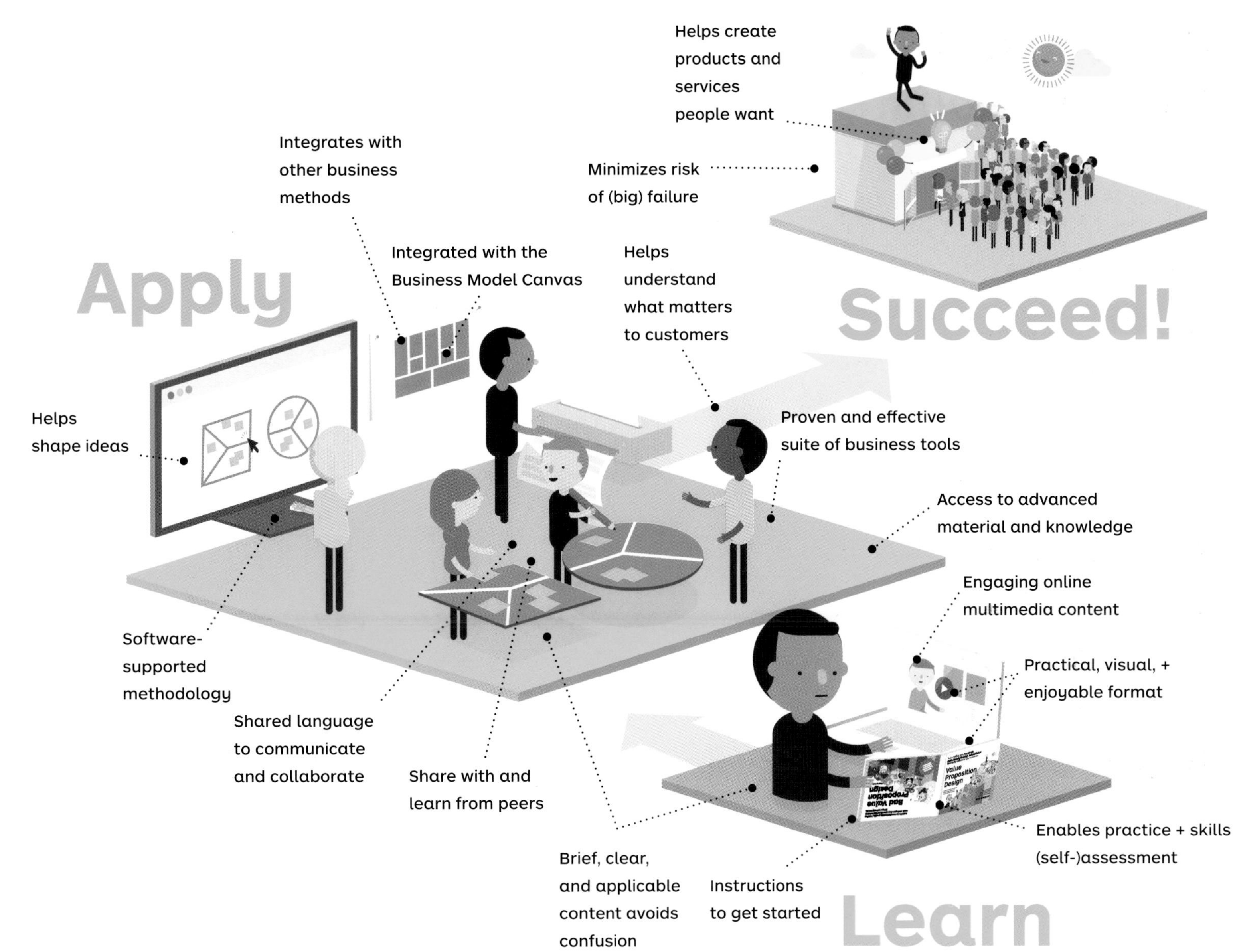
Helps create products and services people want
Minimizes risk of (big) failure
Integrates with other business methods
Integrated with the Business Model Canvas
Helps understand what matters to customers
Apply
Succeed!
Helps shape ideas
Proven and effective suite of business tools
Access to advanced material and knowledge
Engaging online multimedia content
Software-supported methodology
Practical, visual, + enjoyable format
Shared language to communicate and collaborate
Share with and learn from peers
Enables practice + skills (self-)assessment
Brief, clear, and applicable content avoids confusion
Instructions to get started
Learn

The Tools and Process of *Value Proposition Design*

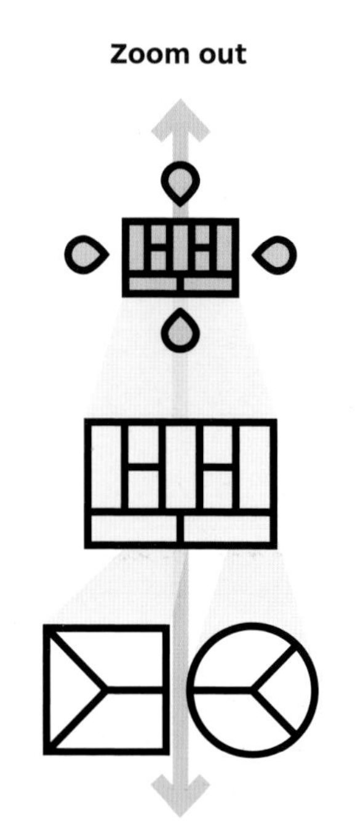

Canvas

Tools

Design / Test

Search

The heart of *Value Proposition Design* is about applying **Tools** to the messy **Search** for value propositions that customers want and then keeping them aligned with what customers want in **Post search**.

Value Proposition Design shows you how to use the **Value Proposition Canvas** to **Design** and **Test** great value propositions in an iterative search for what customers want. Value proposition design is a never-ending process in which you need to **Evolve** your value proposition(s) constantly to keep it relevant to customers.

Progress

Manage the messy and nonlinear process of value proposition design and reduce risk by systematically applying adequate tools and processes.

Evolve

Post search

Design Squiggle adapted from Damien Newman, Central

An Integrated Suite of Tools

The Value Proposition Canvas is the tool at the center of this book. It makes value propositions visible and tangible and thus easier to discuss and manage. It perfectly integrates with the Business Model Canvas and the Environment Map, two tools that are discussed in detail in *Business Model Generation*,* the sister book to this one. Together, they shape the foundation of a suite of business tools.

The Value Proposition Canvas zooms into the details of two of the building blocks of the Business Model Canvas.

* *Business Model Generation*, Osterwalder and Pigneur, 2010.

The
Environment Map
helps you *understand the context in which you create.*

The
Business Model Canvas
helps you
create value for your business.

The
Value Proposition Canvas
helps you
create value for your customer.

Refresher: The Business Model Canvas

Embed your value proposition in a viable business model to capture value for your organization. To do so, you can use the Business Model Canvas, a tool to describe how your organization creates, delivers, and captures value. The Business Model Canvas and Value Proposition Canvas perfectly integrate, with the latter being like a plug-in to the former that allows you to zoom into the details of how you are creating value for customers.

The refresher of the Business Model Canvas on this spread is sufficient to work through this book and create great value propositions. Go to the online resources if you are interested in more or get *Business Model Generation*,* the sister publication to this book.

Customer Segments
are the groups of people and/or organizations a company or organization aims to reach and create value for with a dedicated value proposition.

Value Propositions
are based on a bundle of products and services that create value for a customer segment.

Channels
describe how a value proposition is communicated and delivered to a customer segment through communication, distribution, and sales channels.

Customer Relationships
outline what type of relationship is established and maintained with each customer segment, and they explain how customers are acquired and retained.

Revenue Streams
result from a value proposition successfully offered to a customer segment. It is how an organization captures value with a price that customers are willing to pay.

Key Resources
are the most important assets required to offer and deliver the previously described elements.

Key Activities
are the most important activities an organization needs to perform well.

Key Partnerships
shows the network of suppliers and partners that bring in external resources and activities.

Cost Structure
describes all costs incurred to operate a business model.

Profit
is calculated by subtracting the total of all costs in the cost structure from the total of all revenue streams.

* *Business Model Generation, Osterwalder and Pigneur, 2010.*

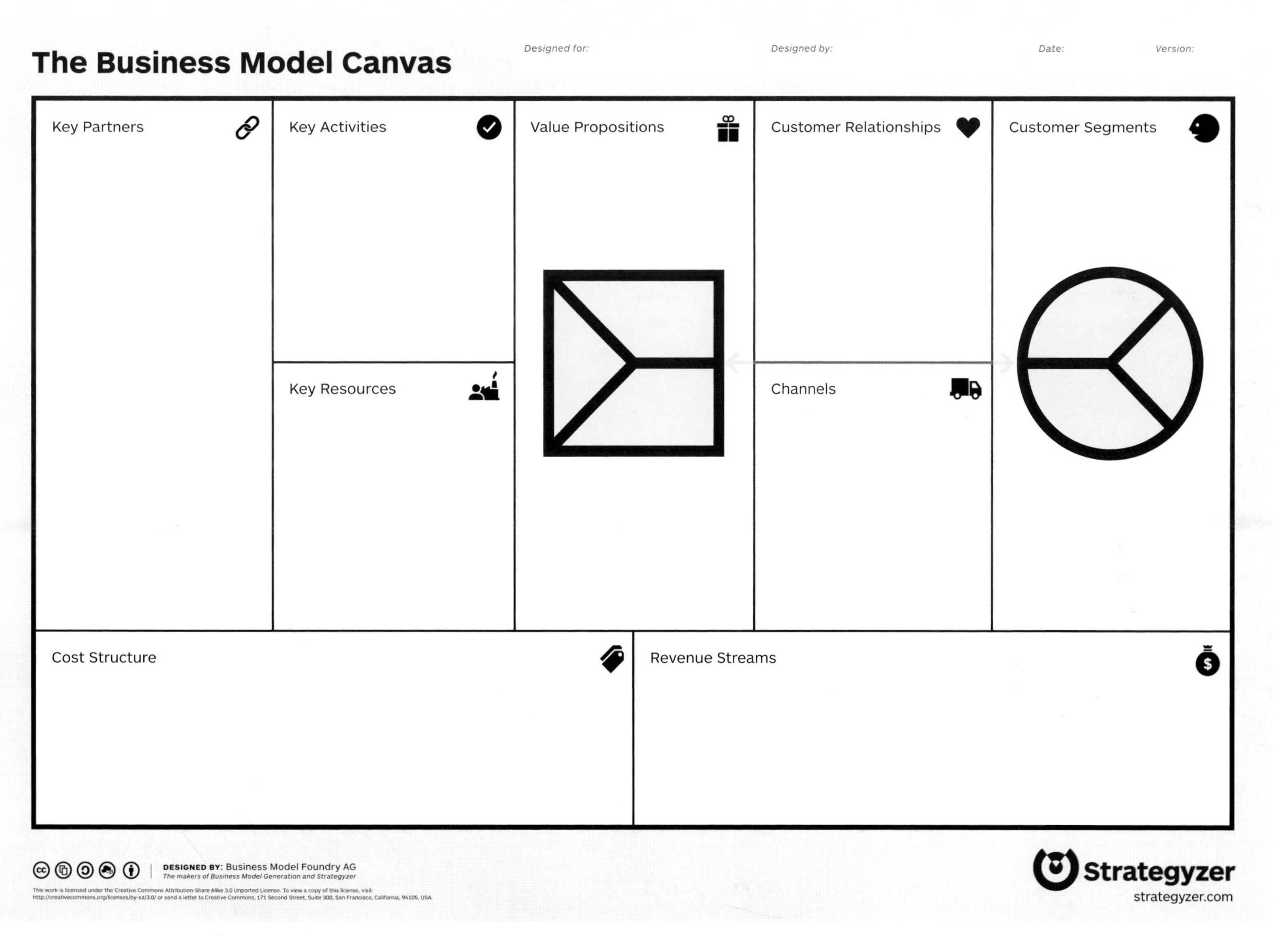

Download detailed Business Model Canvas Explanation and the Business Model Canvas pdf

Value Proposition Design works for...

Are you creating something from scratch on your own or are you part of an existing organization? Some things will be easier and some harder depending on your strategic playground.

A start-up entrepreneur deals with different constraints than a project leader for a new venture within an existing organization. The tools presented in this book apply to both contexts. Depending on your starting point you will execute them in a different way to leverage different strengths and overcome different obstacles.

New Ventures

Individuals or teams setting out to create a great value proposition and business model from scratch

Main challenges

- Produce proof that your ideas can work on a limited budget.
- Manage involvement of investors (if you scale your ideas).
- Risk running out of money before finding the right value proposition and business model.

Main opportunities

- Use speedy decision making and agility to your advantage.
- Leverage the motivation of ownership as a driver for success.

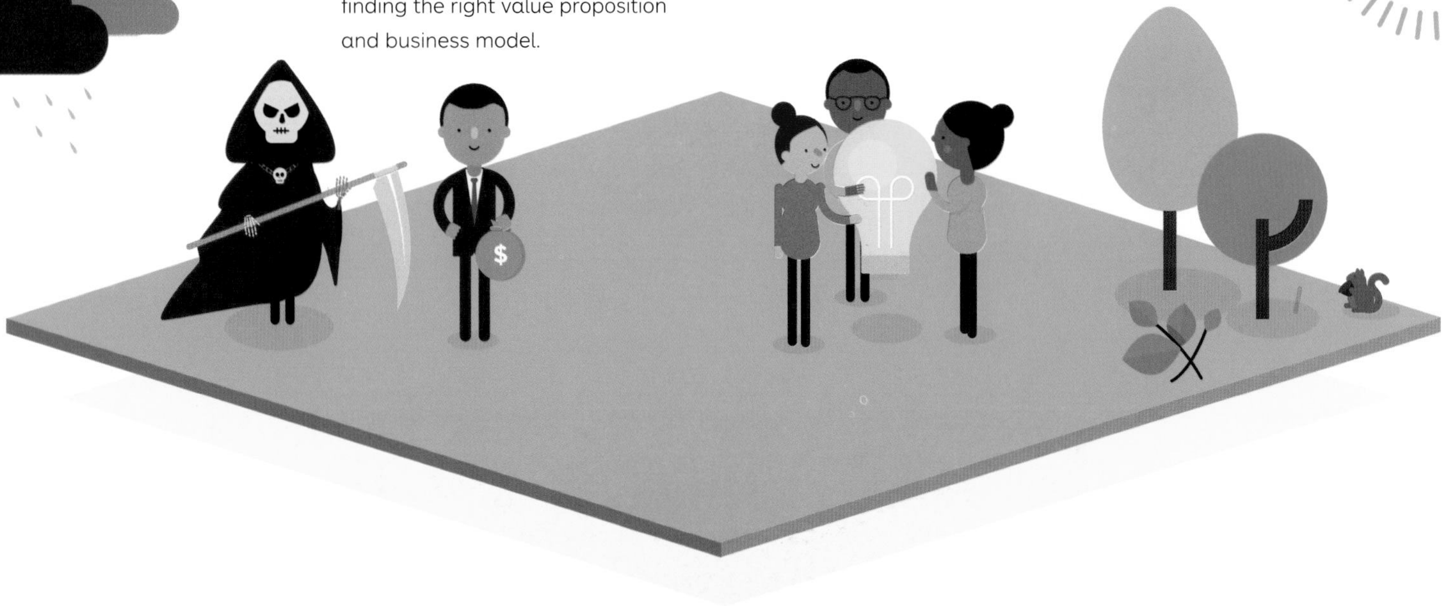

Established Organizations

Teams within existing companies setting out to improve or invent value propositions and business models

Get "Innovating in Established Organizations" poster

Main opportunities

- Build on existing value propositions and business models.
- Leverage existing assets (sales, channels, brand, etc.).
- Build portfolios of business models and value propositions.

Main challenges

- Get buy-in from top management.
- Get access to existing resources.
- Manage cannibalization.
- Overcome risk aversion.
- Overcome rigid and slow processes.
- Produce big wins to move the needle.
- Manage career risk of innovators.

Use *Value Proposition Design* to...

invent and improve value propositions. The tools we will study work for managing and renewing value propositions (and business models) just as much as for creating new ones. Put the value proposition and business model to work to create a shared language of value creation in your organization. Use them to continuously invent and improve value propositions that meet customer profiles, which is an undertaking that never ends.

Invent

Invent new value propositions that people want with business models that work.

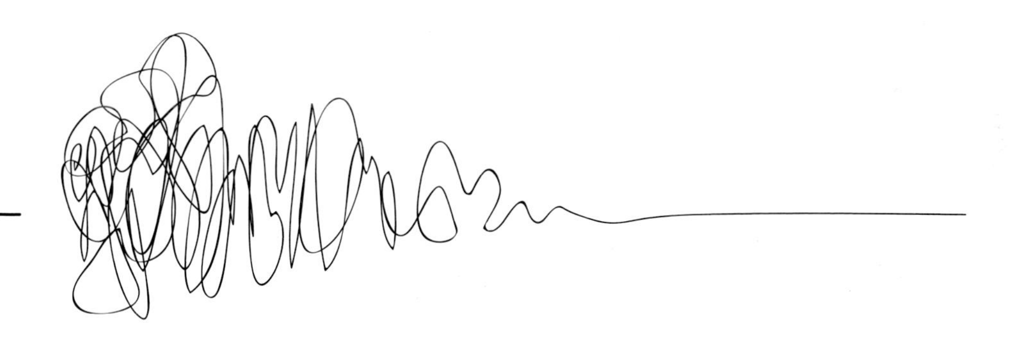

Improve

Manage, measure, challenge, improve, and renew existing value propositions and business models.

Assess Your Value Proposition Design Skills

Complete our online test and assess whether you have the attitude and skills required to systematically be successful at value proposition design. Take the test before and after working through *Value Proposition Design* to measure your progress.

Take your skills test online

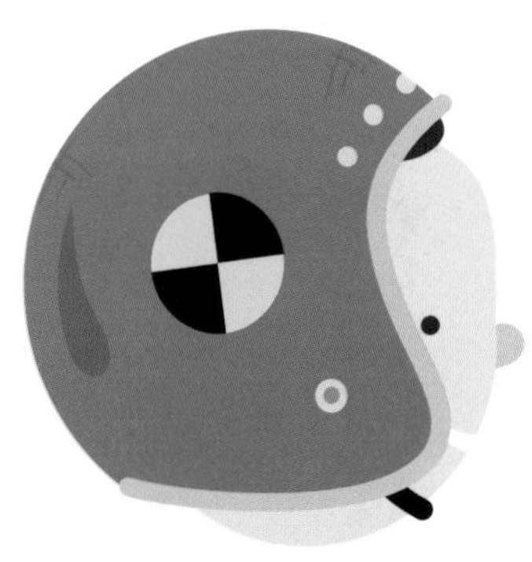

Entrepreneurial Knowledge

You enjoy trying out new things. You don't see the risk of failing as a threat but an opportunity to learn and progress. You easily navigate between the strategic and the tactical.

Tool Skills

You systematically use the Value Proposition Canvas, Business Model Canvas, and other tools and processes in your search for great value propositions and business models.

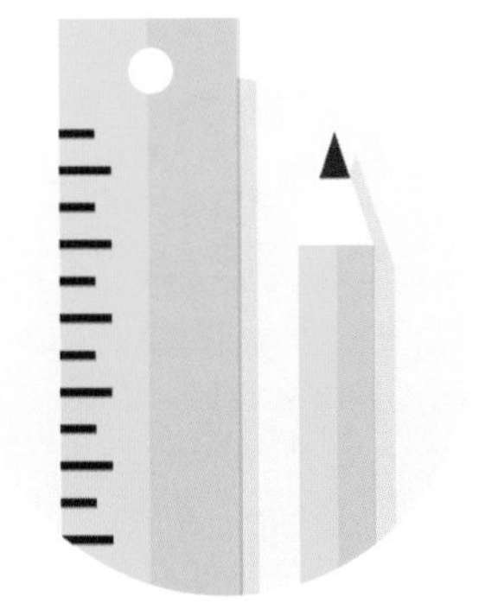

Design Thinking Skills

You explore multiple alternatives before picking and refining a particular direction. You are comfortable with the nonlinear and iterative nature of value creation.

Customer Empathy

You relentlessly take a customer perspective and are even better at listening to customers than selling to them.

Experimentation Skills

You systematically seek evidence that supports your ideas and tests your vision. You experiment at the earliest stages to learn what works and what doesn't.

Sell Your Colleagues on Value Proposition Design

concerned that we don't have a methodology to track our progress on the development of that new value proposition and business model.

I am...

worried that we focus too much on products and features instead of creating value for customers.

astonished at how poorly aligned product development, sales, and marketing are when it comes to developing new value propositions.

surprised at how often we make stuff nobody wants, despite our good ideas and good intentions.

really disappointed by how much we talked about value propositions and business models at our last meeting without really getting tangible results.

blown away by how unclear that last presentation on that new value proposition and business model was.

amazed by how many resources we wasted when that great idea in that last business plan turned out to be a flop because we didn't test it.

concerned that our product development process doesn't use a more customer-focused methodology.

surprised that we invest so much in research and development (R&D), but fail to invest in developing the right value propositions and business models.

not sure if everybody in our team has a shared understanding of what a good value proposition actually is.

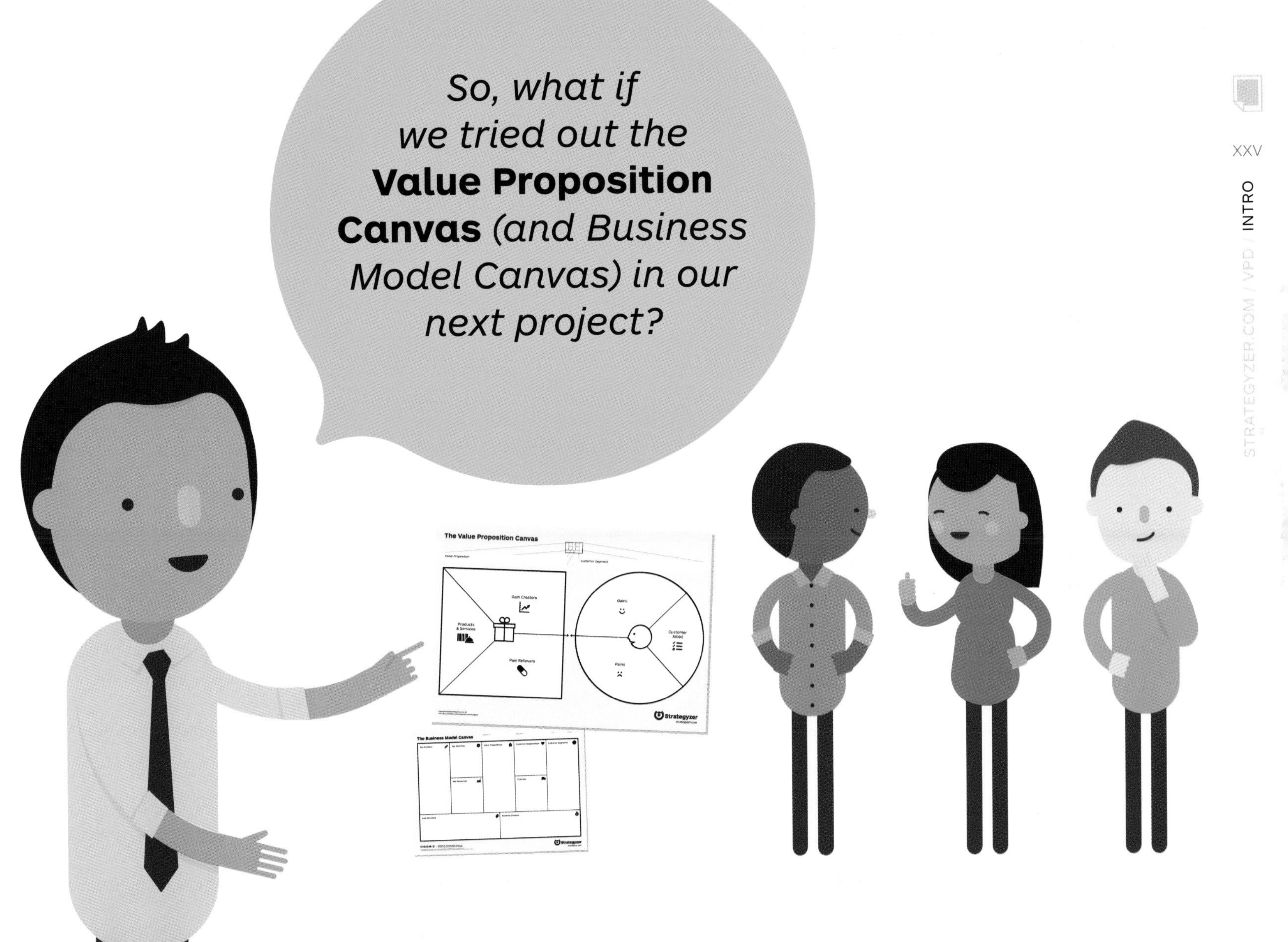

Get a slide deck with 10 arguments to use with the Value Proposition and Business Model Canvases

can

vas

1

The Value Proposition Canvas has two sides. With the **Customer Profile** p. 10 you clarify your customer understanding. With the **Value Map** p. 26 you describe how you intend to create value for that customer. You achieve **Fit** p. 40 between the two when one meets the other.

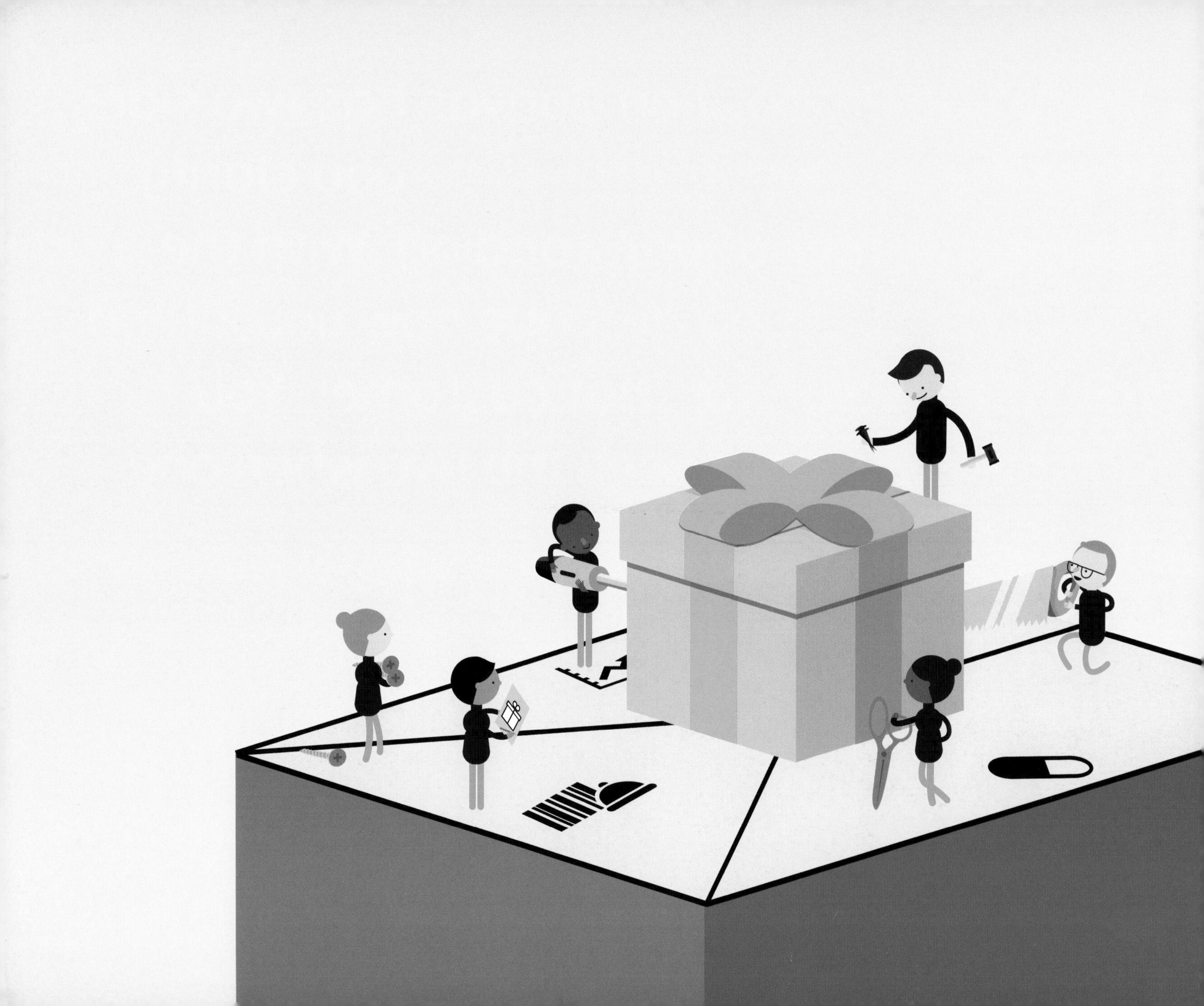

F

Create Value

The set of value proposition **benefits** that you **design** to attract customers.

DEF·I·NI·TION

VALUE PROPOSITION

Describes the benefits customers can expect from your products and services.

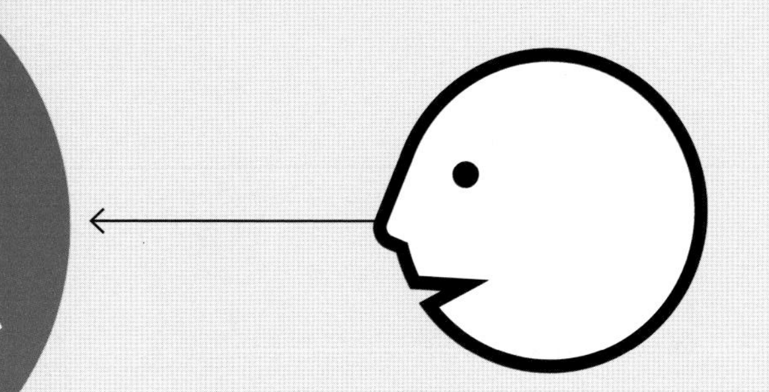

Observe Customers

The set of customer **characteristics** that you **assume, observe, and verify** in the market.

Value Map

The Value (Proposition) Map describes the features of a specific value proposition in your business model in a more structured and detailed way. It breaks your value proposition down into products and services, pain relievers, and gain creators.

Gain Creators describe how your products and services create customer gains.

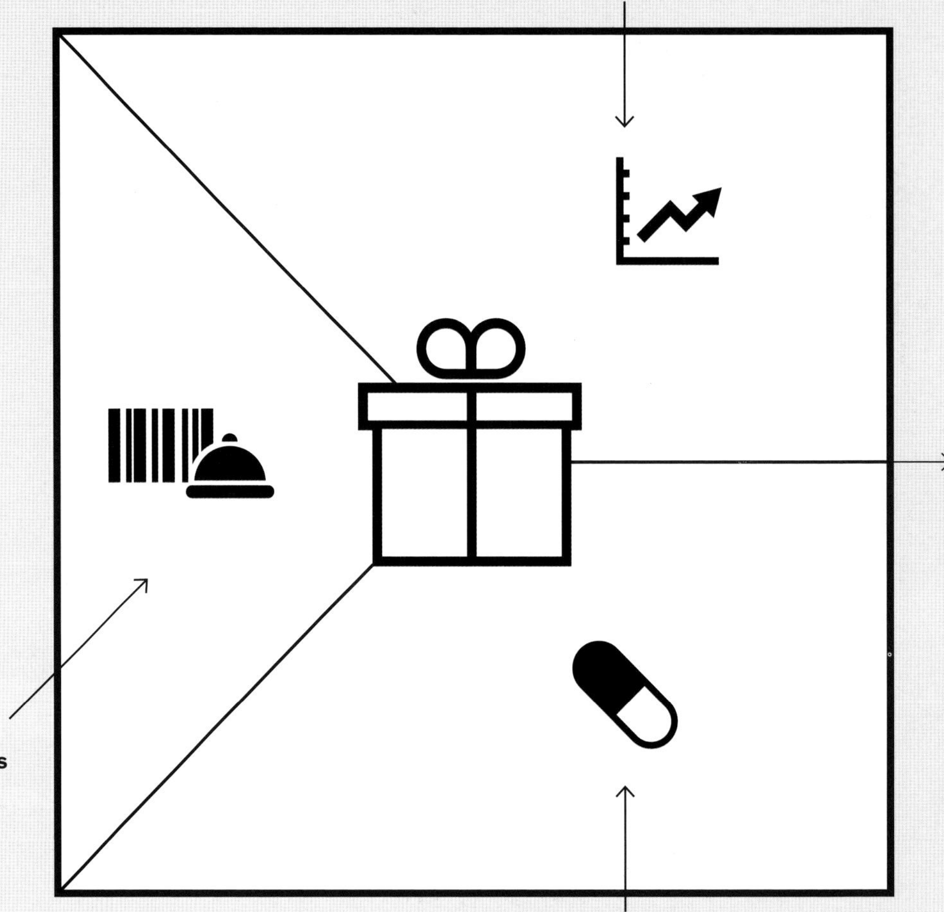

This is a list of all the **Products and Services** a value proposition is built around.

Pain Relievers describe how your products and services alleviate customer pains.

Customer Profile

The Customer (Segment) Profile describes a specific customer segment in your business model in a more structured and detailed way. It breaks the customer down into its jobs, pains, and gains.

Gains describe the outcomes customers want to achieve or the concrete benefits they are seeking.

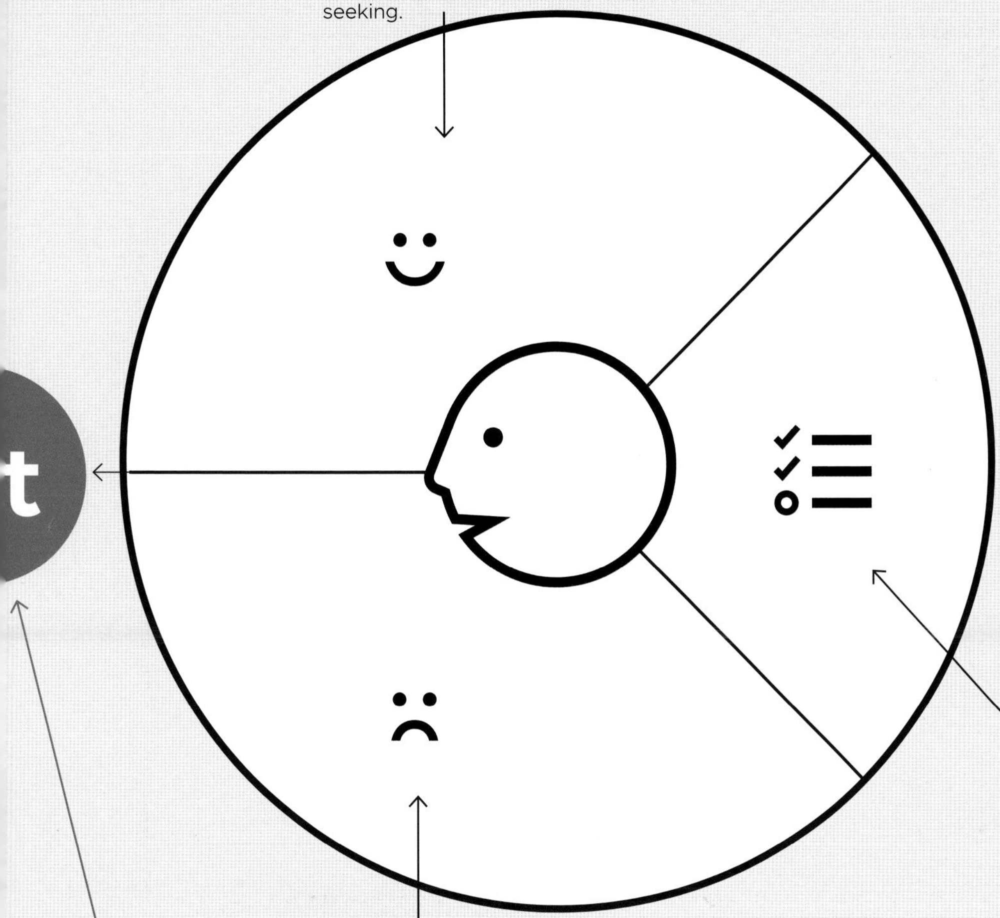

Customer Jobs describe what customers are trying to get done in their work and in their lives, as expressed in their own words.

Pains describe bad outcomes, risks, and obstacles related to customer jobs.

You achieve **Fit** when your value map meets your customer profile—when your products and services produce pain relievers and gain creators that match one or more of the jobs, pains, and gains that are important to your customer.

t

1.1

Customer Profile

Customer Jobs

Jobs describe the things your customers are trying to get done in their work or in their life. A customer job could be the tasks they are trying to perform and complete, the problems they are trying to solve, or the needs they are trying to satisfy. Make sure you take the customer's perspective when investigating jobs. What you think of as important from your perspective might not be a job customers are actually trying to get done.*

Distinguish between three main types of customer jobs to be done and supporting jobs:

Functional jobs
When your customers try to perform or complete a specific task or solve a specific problem, for example, mow the lawn, eat healthy as a consumer, write a report, or help clients as a professional.

Social jobs
When your customers want to look good or gain power or status. These jobs describe how customers want to be perceived by others, for example, look trendy as a consumer or be perceived as competent as a professional.

Personal/emotional jobs
When your customers seek a specific emotional state, such as feeling good or secure, for example, seeking peace of mind regarding one's investments as a consumer or achieving the feeling of job security at one's workplace.

Supporting jobs
Customers also perform supporting jobs in the context of purchasing and consuming value either as consumers or as professionals. These jobs arise from three different roles:

- BUYER OF VALUE: jobs related to buying value, such as comparing offers, deciding which products to buy, standing in a checkout line, completing a purchase, or taking delivery of a product or service.

- COCREATOR OF VALUE: jobs related to cocreating value with your organization, such as posting product reviews and feedback or even participating in the design of a product or service.

- TRANSFERRER OF VALUE: jobs related to the end of a value proposition's life cycle, such as canceling a subscription, disposing of a product, transferring it to others, or reselling it.

* *The jobs to be done concept was developed independently by several business thinkers including Anthony Ulwick of the consulting firm* Strategyn, *consultants Rick Pedi and Bob Moesta, and Professor Denise Nitterhouse of DePaul University. It was popularized by Clay Christensen and his consulting firm* Innosight *and Anthony Ulwick's* Strategyn.

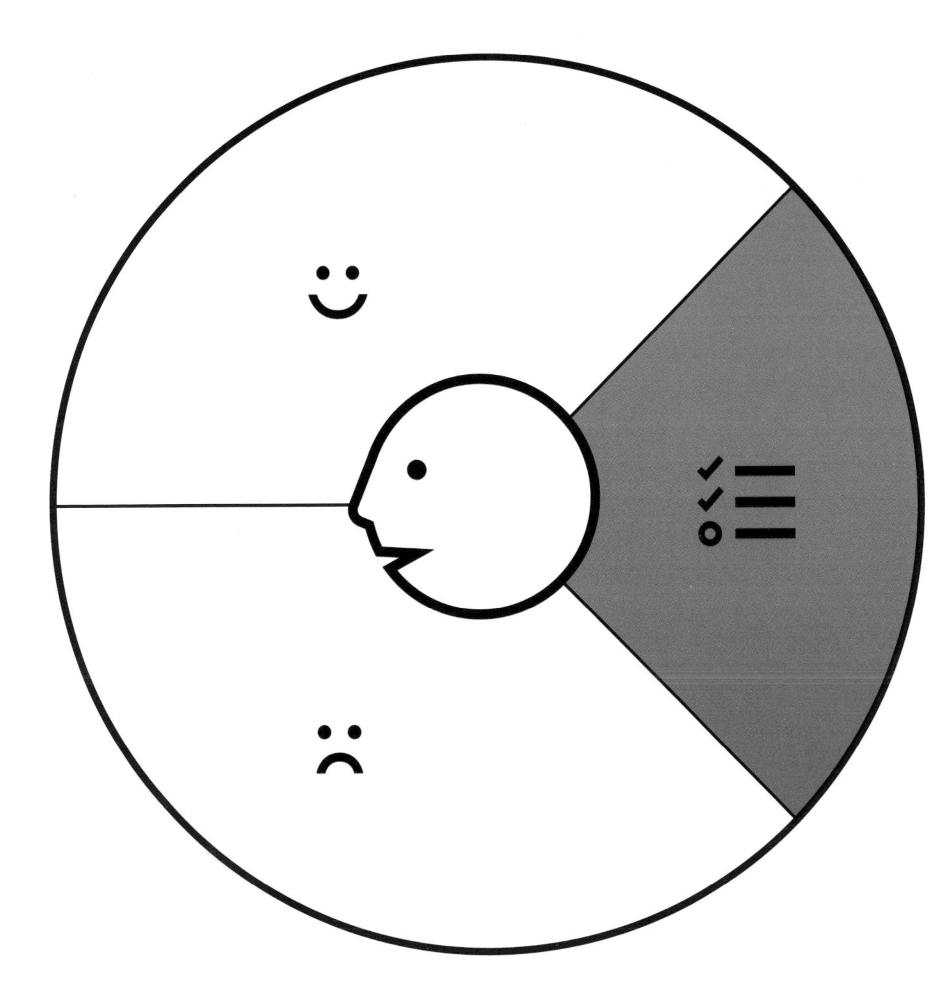

Job context

Customer jobs often depend on the specific context in which they are performed. The context may impose certain constraints or limitations. For example, calling somebody on the fly is different when you are traveling on a train than when you are driving a car. Likewise, going to the movies with your kids is different than going with your partner.

Job importance

It is important to acknowledge that not all jobs have the same importance to your customer. Some matter more in a customer's work or life because failing to get them done could have serious ramifications. Some are insignificant because the customer cares about other things more. Sometimes a customer will deem a job crucial because it occurs frequently or because it will result in a desired or unwanted outcome.

+
Important
↕
Insignificant
−

Download trigger questions to help find customer jobs

Customer Pains

Pains describe anything that annoys your customers before, during, and after trying to get a job done or simply prevents them from getting a job done. Pains also describe risks, that is, potential bad outcomes, related to getting a job done badly or not at all.

Seek to identify three types of customer pains and how severe customers find them:

Undesired outcomes, problems, and characteristics
Pains are functional (e.g., a solution doesn't work, doesn't work well, or has negative side effects), social ("I look bad doing this"), emotional ("I feel bad every time I do this"), or ancillary ("It's annoying to go to the store for this"). This may also involve undesired characteristics customers don't like (e.g., "Running at the gym is boring," or "This design is ugly").

Obstacles
These are things that prevent customers from even getting started with a job or that slow them down (e.g., "I lack the time to get this job done accurately," or "I can't afford any of the existing solutions").

Risks (undesired potential outcomes)
What could go wrong and have important negative consequences (e.g., "I might lose credibility when using this type of solution," or "A security breach would be disastrous for us").

Pain severity

A customer pain can be extreme or moderate, similar to how jobs can be important or insignificant to the customer.

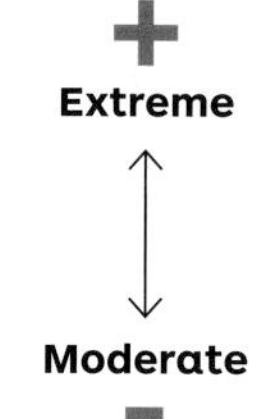

Tip: Make pains concrete.
To clearly differentiate jobs, pains, and gains, describe them as concretely as possible. For example, when a customer says "waiting in line was a waste of time," ask after how many minutes exactly it began to feel like wasted time. That way you can note "wasting more than *x* minutes standing in line." When you understand how exactly customers measure pain severity, you can design better pain relievers in your value proposition.

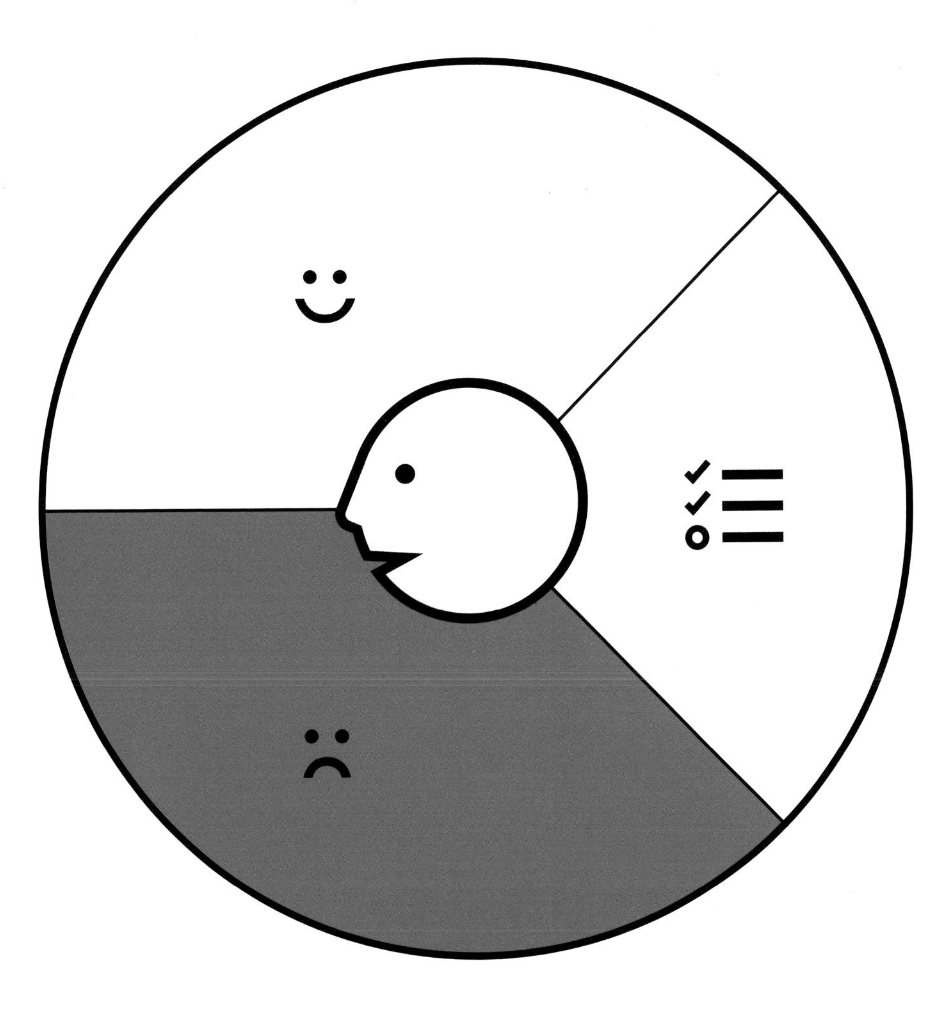

The following list of trigger questions can help you think of different potential customer pains:

- How do your customers define too costly? Takes a lot of time, costs too much money, or requires substantial efforts?
- What makes your customers feel bad? What are their frustrations, annoyances, or things that give them a headache?
- How are current value propositions underperforming for your customers? Which features are they missing? Are there performance issues that annoy them or malfunctions they cite?
- What are the main difficulties and challenges your customers encounter? Do they understand how things work, have difficulties getting certain things done, or resist particular jobs for specific reasons?
- What negative social consequences do your customers encounter or fear? Are they afraid of a loss of face, power, trust, or status?
- What risks do your customers fear? Are they afraid of financial, social, or technical risks, or are they asking themselves what could go wrong?
- What's keeping your customers awake at night? What are their big issues, concerns, and worries?
- What common mistakes do your customers make? Are they using a solution the wrong way?
- What barriers are keeping your customers from adopting a value proposition? Are there upfront investment costs, a steep learning curve, or other obstacles preventing adoption?

Download trigger questions

Customer Gains

Gains describe the outcomes and benefits your customers want. Some gains are required, expected, or desired by customers, and some would surprise them. Gains include functional utility, social gains, positive emotions, and cost savings.

Seek to identify four types of customer gains in terms of outcomes and benefits:

Required gains
These are gains without which a solution wouldn't work. For example, the most basic expectation that we have from a smartphone is that we can make a call with it.

Expected gains
These are relatively basic gains that we expect from a solution, even if it could work without them. For example, since Apple launched the iPhone, we expect phones to be well-designed and look good.

Desired gains
These are gains that go beyond what we expect from a solution but would love to have if we could. These are usually gains that customers would come up with if you asked them. For example, we desire smartphones to be seamlessly integrated with our other devices.

Unexpected gains
These are gains that go beyond customer expectations and desires. They wouldn't even come up with them if you asked them. Before Apple brought touch screens and the App Store to the mainstream, nobody really thought of them as part of a phone.

Gain relevance
A customer gain can feel essential or nice to have, just like pains can feel extreme or moderate to them.

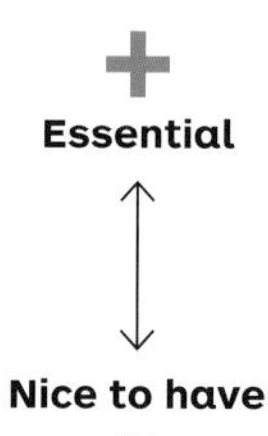

Tip: Make gains concrete.
As with pains, it's better to describe gains as concretely as possible to clearly differentiate jobs, pains, and gains from one another. Ask how much they'd expect or dream of when a customer indicates "better performance" as a desired gain. That way you can note "would love an increased performance of more than *x*." When you understand how exactly customers measure gains (i.e., outcomes and benefits), you can design better gain creators in your value proposition.

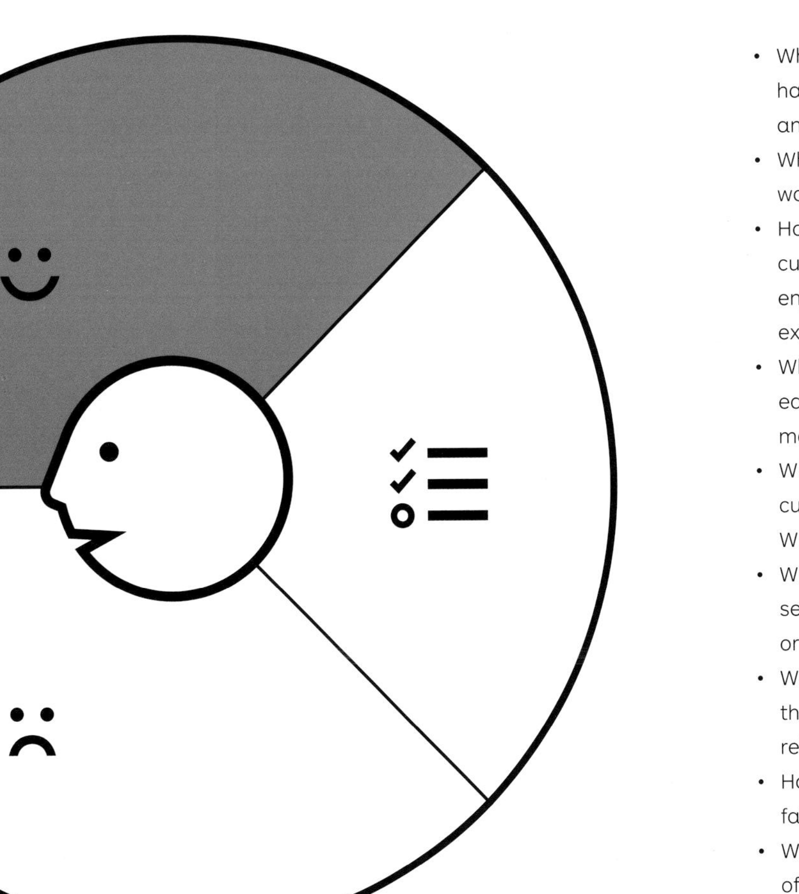

The following list of trigger questions can help you think of different potential customer gains:

- Which savings would make your customers happy? Which savings in terms of time, money, and effort would they value?
- What quality levels do they expect, and what would they wish for more or less of?
- How do current value propositions delight your customers? Which specific features do they enjoy? What performance and quality do they expect?
- What would make your customers' jobs or lives easier? Could there be a flatter learning curve, more services, or lower costs of ownership?
- What positive social consequences do your customers desire? What makes them look good? What increases their power or their status?
- What are customers looking for most? Are they searching for good design, guarantees, specific or more features?
- What do customers dream about? What do they aspire to achieve, or what would be a big relief to them?
- How do your customers measure success and failure? How do they gauge performance or cost?
- What would increase your customers' likelihood of adopting a value proposition? Do they desire lower cost, less investment, lower risk, or better quality?

Download trigger questions

Profile of a "Business Book Reader"

We chose to use potential readers of this book to illustrate the customer profile. We deliberately went beyond jobs, pains, and gains merely related to reading books, since we intended to design an innovative and more holistic value proposition for businesspeople in general.

The customer profile sketched out on the right is informed by several interviews we conducted and thousands of interactions we had with workshop participants. However, it is not mandatory to start with preexisting customer knowledge. You may begin exploring ideas by sketching out a profile based on what you believe your potential customers look like. This is an excellent starting point to prepare customer interviews and tests regarding your assumptions about customer jobs, pains, and gains.

Gains are benefits, results, and characteristics that customers require or desire. They are outcomes of jobs or wanted characteristics of a value proposition that help customers get a job done well.

The more tangible and specific you make pains and gains, the better. For example, "examples from my industry" is more concrete than "relevant to my context." Ask customers how they measure gains and pains. Investigate how they measure success or failure of a job they want to get done.

Make sure you deeply understand your customer. If you have only a few sticky notes on your profile, that probably indicates a lack of customer understanding. Unearth as many jobs, pains, and gains as you can. Search beyond those directly related to your value proposition.

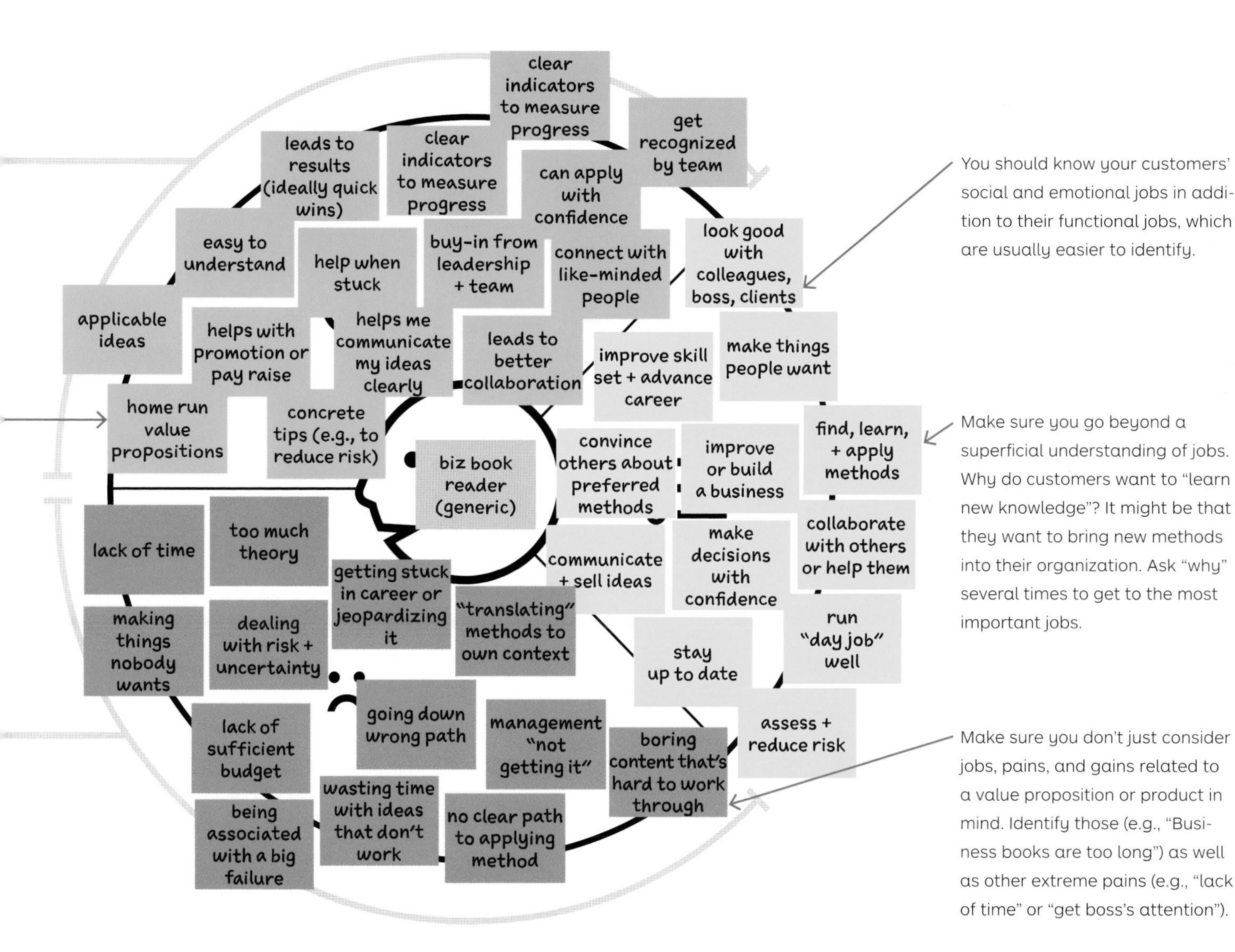

You should know your customers' social and emotional jobs in addition to their functional jobs, which are usually easier to identify.

Make sure you go beyond a superficial understanding of jobs. Why do customers want to "learn new knowledge"? It might be that they want to bring new methods into their organization. Ask "why" several times to get to the most important jobs.

Make sure you don't just consider jobs, pains, and gains related to a value proposition or product in mind. Identify those (e.g., "Business books are too long") as well as other extreme pains (e.g., "lack of time" or "get boss's attention").

Ranking Jobs, Pains, and Gains

Although individual customer preferences vary, you need to get a sense of customer priorities. Investigate which jobs the majority consider important or insignificant. Find out which pains they find extreme versus merely moderate. Learn which gains they find essential and which are simply nice to have.

Ranking jobs, pains, and gains is essential in order to design value propositions that address things customers really care about. Of course, it's difficult to unearth what really matters to customers, but your understanding will improve with every customer interaction and experiment.

It doesn't matter if you start out with a ranking that is based on what you think is important to your potential customers as long as you strive to test that ranking until it truly reflects priorities from the customer's perspective.

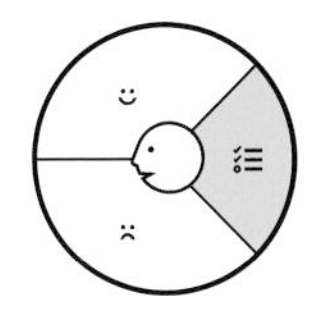

Job importance
Rank jobs according to their importance to customers.

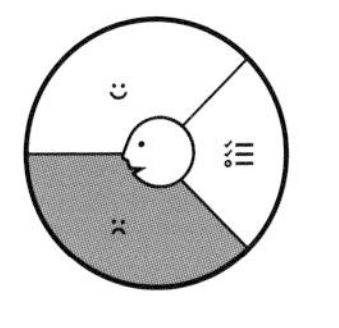

Pain severity
Rank pains according to how extreme they are in the customers' eyes.

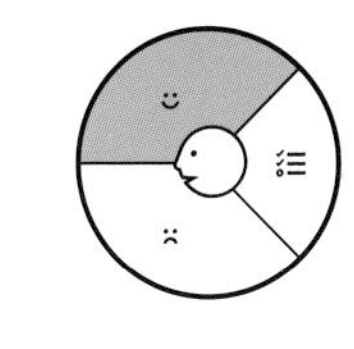

Gain relevance
Rank gains according to how essential they are in the customers' eyes.

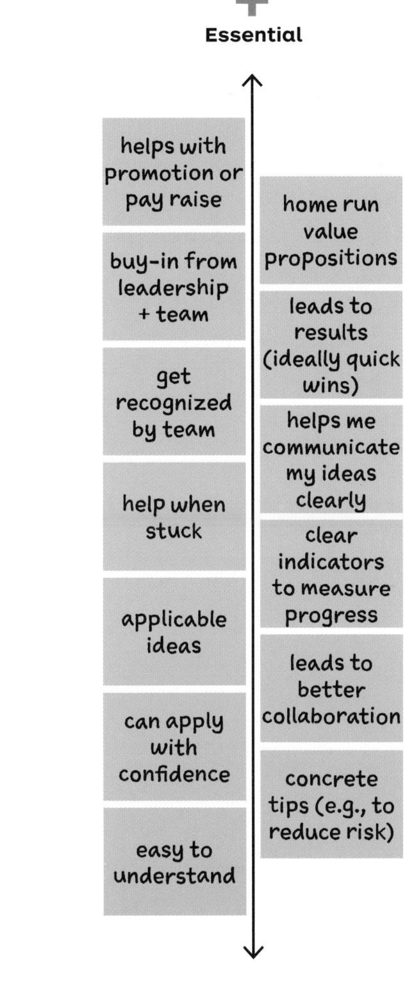

Step into Your Customers' Shoes

OBJECTIVE
Visualize what matters to your customers in a shareable format

OUTCOME
One page actionable customer profile

How good is your understanding of your customers' jobs, pains, and gains? Map out a customer profile.

Instructions

Map the profile of one of your currently existing customer segments to practice using the customer profile. If you are working on a new idea, sketch out the customer segment you intend to create value for.

1. Download the Customer Profile Canvas.
2. Grab a set of small sticky notes.
3. Map out your customer profile.

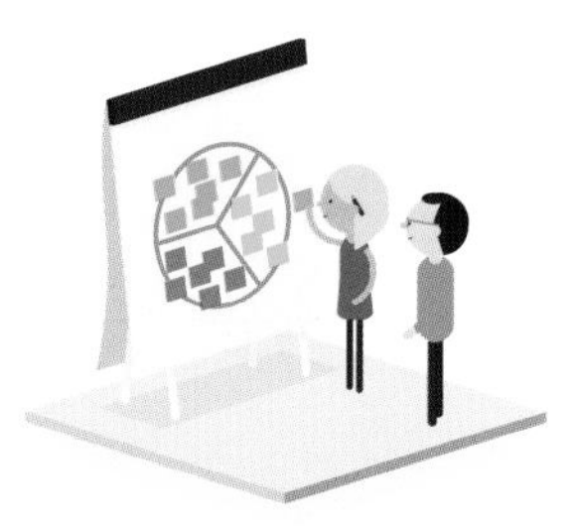

1

Select customer segment.
Select a customer segment that you want to profile.

2

Identify customer jobs.
Ask what tasks your customers are trying to complete. Map out all of their jobs by writing each one on an individual sticky note.

3

Identify customer pains.
What pains do your customers have? Write down as many as you can come up with, including obstacles and risks.

4

Identify customer gains.
What outcomes and benefits do your customers want to achieve? Write down as many gains as you can come up with.

5

Prioritize jobs, pains, and gains.
Order jobs, pains, and gains in a column, each with the most important jobs, most extreme pains, and essential gains on top and the moderate pains and nice-to-have gains at the bottom.

Do this exercise online

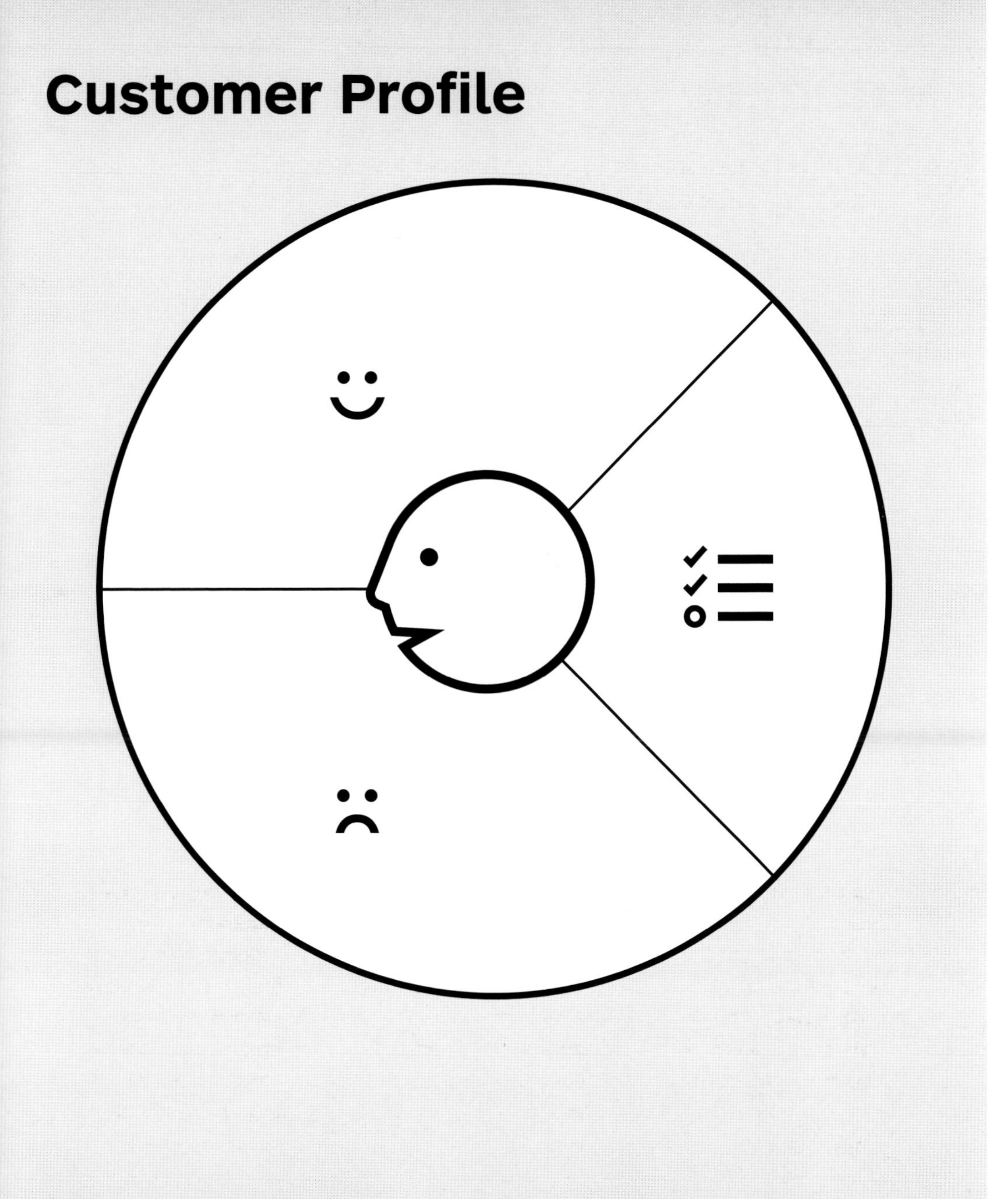

Download the Customer Profile pdf

Best Practices for Mapping Jobs, Pains, and Gains

Avoid frequently committed mistakes when profiling a customer, and instead follow these best practices.

✗ Common Mistakes

Mixing several customer segments into one profile.

Mixing jobs and outcomes.

Focusing on functional jobs only and forgetting social and emotional jobs.

Listing jobs, pains, and gains with your value proposition in mind.

Identifying too few jobs, pains, and gains.

Being too vague in descriptions of pains and gains.

✓ Best Practices

Make a Value Proposition Canvas for every different customer segment. If you sell to companies, ask yourself if you have different types of customers within each company (e.g., users, buyers).

Jobs are the tasks customers are trying to perform, the problems they are trying to solve, or the needs they are trying to satisfy, whereas gains are the concrete outcomes they want to achieve—or avoid and eliminate in the case of pains.

Sometimes social or emotional jobs are even more important than the "visible" functional jobs. "Looking good in front of others" might be more important than finding a great technical solution that helps complete the job effectively.

When you map your customer, you should proceed like an anthropologist and "forget" what you are offering. For example, a business publisher should not map jobs, pains, and gains merely related to books, because a reader has the choice between business books, consultants, YouTube videos, or even completing an MBA program or training. Go beyond the jobs, pains, and gains you intend or hope to address with your value proposition.

A good customer profile is full of sticky notes, because most customers have a lot of pains and expect or desire a lot of gains. Map out all your (potential) customers' important jobs, extreme pains, and essential gains.

Make pains and gains tangible and concrete. Rather than just writing "salary increase" in gains, specify how much of an increase a customer is seeking. Rather than writing "takes too long" in pains, indicate how long "too long" actually is. This will allow you to understand how exactly customers measure success and failure.

Pains vs. Gains

When you get started with the customer profile, you might simply put the same ideas in pains and gains as opposites of each other. For example, if one of the customers' jobs to be done is "earn more money," you might start by adding "salary increase" to gains and "salary decrease" to pains.

Here's a better way to do it:

- Find out precisely how much more money the customer expects to earn so it feels like a gain and investigate what decrease would feel like a pain.
- In the pains, add the barriers that prevent or make it difficult to get a job done. In our example the pain might be "my employer doesn't give raises."
- In the pains, add the risks related to not getting the job done. In our example the pain could be "might not be able to afford my child's future college tuition."

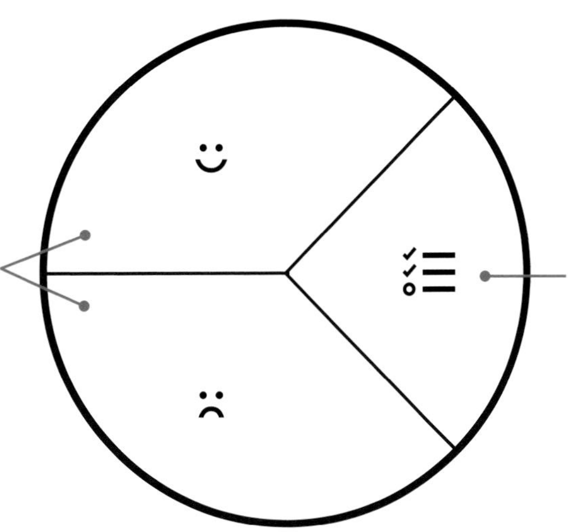

Ask "why" several times until you really understand your customers' jobs to be done.

Another issue when you get started with the customer profile is that you might settle with a superficial understanding of your customer's jobs. To avoid this, you need to ask yourself why a customer wants to perform a certain job to dig deeper toward the real motivations.

For example, why might a customer want to learn a foreign language? Maybe because the "real" customer job to be done is to improve his CV. Why does he want to improve his CV? Maybe because he wants to earn more money.

Don't settle until you really understand the underlying jobs to be done that really drive customers.

26

1.2

Value Map

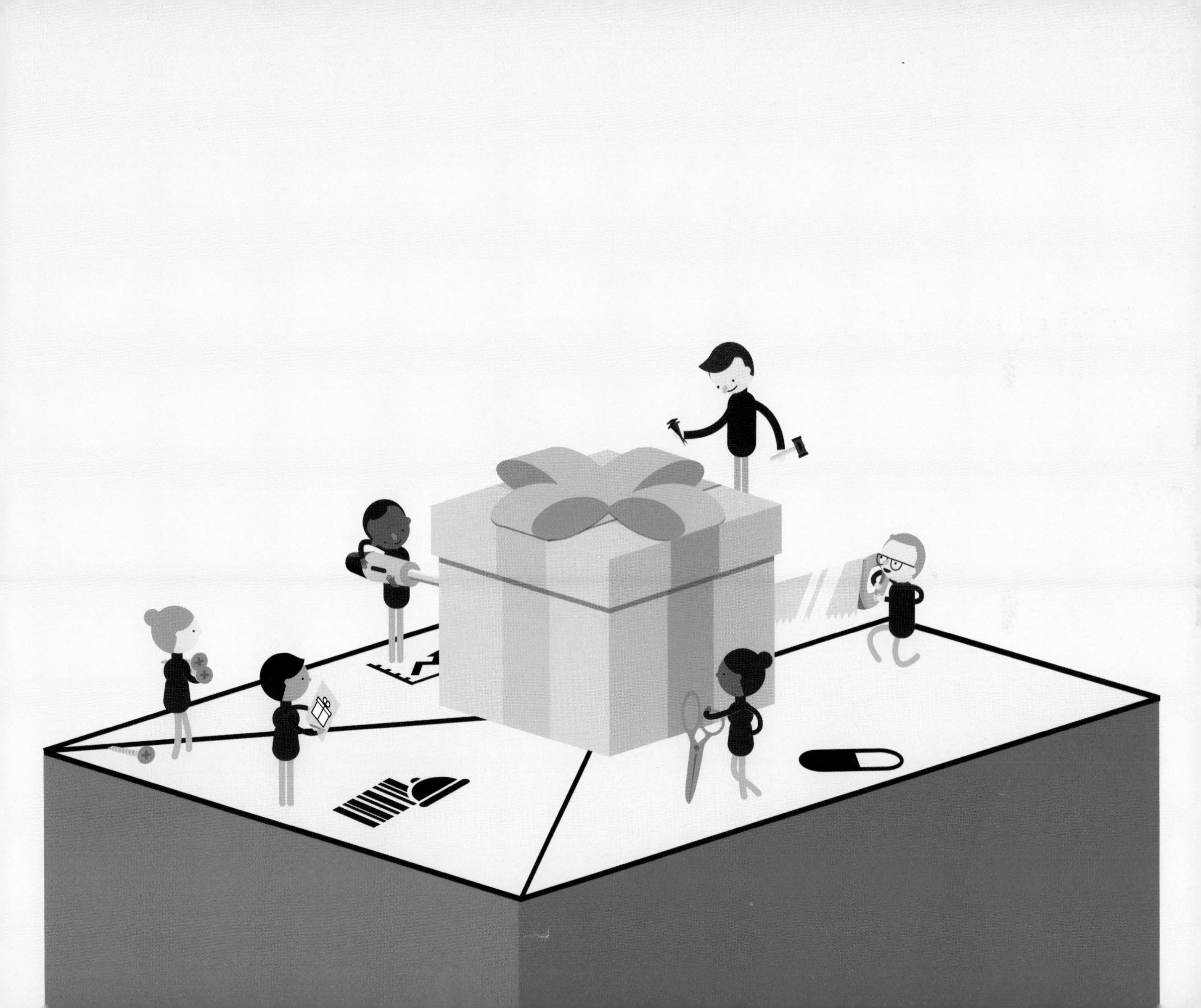

Products and Services

This is simply a list of what you offer. Think of it as all the items your customers can see in your shop window—metaphorically speaking. It's an enumeration of all the products and services your value proposition builds on. This bundle of products and services helps your customers complete either functional, social, or emotional jobs or helps them satisfy basic needs. It is crucial to acknowledge that products and services don't create value alone—only in relationship to a specific customer segment and their jobs, pains, and gains.

Your list of products and services may also include supporting ones that help your customers perform the roles of buyer (those that help customers compare offers, decide, and buy), co-creator (those that help customers co-design value propositions), and transferrer (those that help customers dispose of a product).

Your value proposition is likely to be composed of various types of products and services:

Physical/tangible
Goods, such as manufactured products.

Intangible
Products such as copyrights or services such as after-sales assistance.

Digital
Products such as music downloads or services such as online recommendations.

Financial
Products such as investment funds and insurances or services such as the financing of a purchase.

Relevance
It is essential to acknowledge that not all products and services have the same relevance to your customers. Some products and services are essential to your value proposition; some are merely nice to have.

Pain Relievers

Pain relievers describe how exactly your products and services alleviate specific customer pains. They explicitly outline how you intend to eliminate or reduce some of the things that annoy your customers before, during, or after they are trying to complete a job or that prevent them from doing so.

Great value propositions focus on pains that matter to customers, in particular extreme pains. You don't need to come up with a pain reliever for every pain you've identified in the customer profile—no value proposition can do this. Great value propositions often focus only on few pains that they alleviate extremely well.

The following list of trigger questions can help you think of different ways your products and services may help your customers alleviate pains.

Ask yourself: Could your products and services...

- produce savings? In terms of time, money, or efforts.
- make your customers feel better? By killing frustrations, annoyances, and other things that give customers a headache.
- fix underperforming solutions? By introducing new features, better performance, or enhanced quality.
- put an end to difficulties and challenges your customers encounter? By making things easier or eliminating obstacles.
- wipe out negative social consequences your customers encounter or fear? In terms of loss of face or lost power, trust, or status.
- eliminate risks your customers fear? In terms of financial, social, technical risks, or things that could potentially go wrong.
- help your customers better sleep at night? By addressing significant issues, diminishing concerns, or eliminating worries.
- limit or eradicate common mistakes customers make? By helping them use a solution the right way.
- eliminate barriers that are keeping your customer from adopting value propositions? Introducing lower or no upfront investment costs, a flatter learning curve, or eliminating other obstacles preventing adoption.

Relevance

A pain reliever can be more or less valuable to the customer. Make sure you differentiate between essential pain relievers and ones that are nice to have. The former relieve extreme issues, often in a radical way, and create a lot of value. The latter merely relieve moderate pains.

+
Essential
↕
Nice to have
−

Download trigger questions

Gain Creators

Gain creators describe how your products and services create customer gains. They explicitly outline how you intend to produce outcomes and benefits that your customer expects, desires, or would be surprised by, including functional utility, social gains, positive emotions, and cost savings.

As with pain relievers, gain creators don't need to address every gain identified in the customer profile. Focus on those that are relevant to customers and where your products and services can make a difference.

The following list of trigger questions can help you think of different ways your products and services may help your customers obtain required, expected, desired, or unexpected outcomes and benefits.

Ask yourself: Could your products and services...

- create savings that please your customers? In terms of time, money, and effort.
- produce outcomes your customers expect or that exceed their expectations? By offering quality levels, more of something, or less of something.
- outperform current value propositions and delight your customers? Regarding specific features, performance, or quality.
- make your customers' work or life easier? Via better usability, accessibility, more services, or lower cost of ownership.
- create positive social consequences? By making them look good or producing an increase in power or status.
- do something specific that customers are looking for? In terms of good design, guarantees, or specific or more features.
- fulfill a desire customers dream about? By helping them achieve their aspirations or getting relief from a hardship?
- produce positive outcomes matching your customers' success and failure criteria? In terms of better performance or lower cost.
- help make adoption easier? Through lower cost, fewer investments, lower risk, better quality, improved performance, or better design.

Relevance

A gain creator can produce more or less relevant outcomes and benefits for the customer just like we have seen for pain relievers. Make sure you differentiate between essential and nice to have gain creators.

+ Essential
↕
Nice to have −

Download trigger questions

Mapping the Value Proposition of Value Proposition Design

Remarkable value propositions focus on jobs, pains, and gains that matter to customers and achieve those exceedingly well. Again, you should not try to address all customer pains and gains. Focus on those that will make a difference for your customer.

It's okay to aggregate several value propositions into one.

"Naked" list of the products and services that your value proposition builds on to target a specific customer segment.

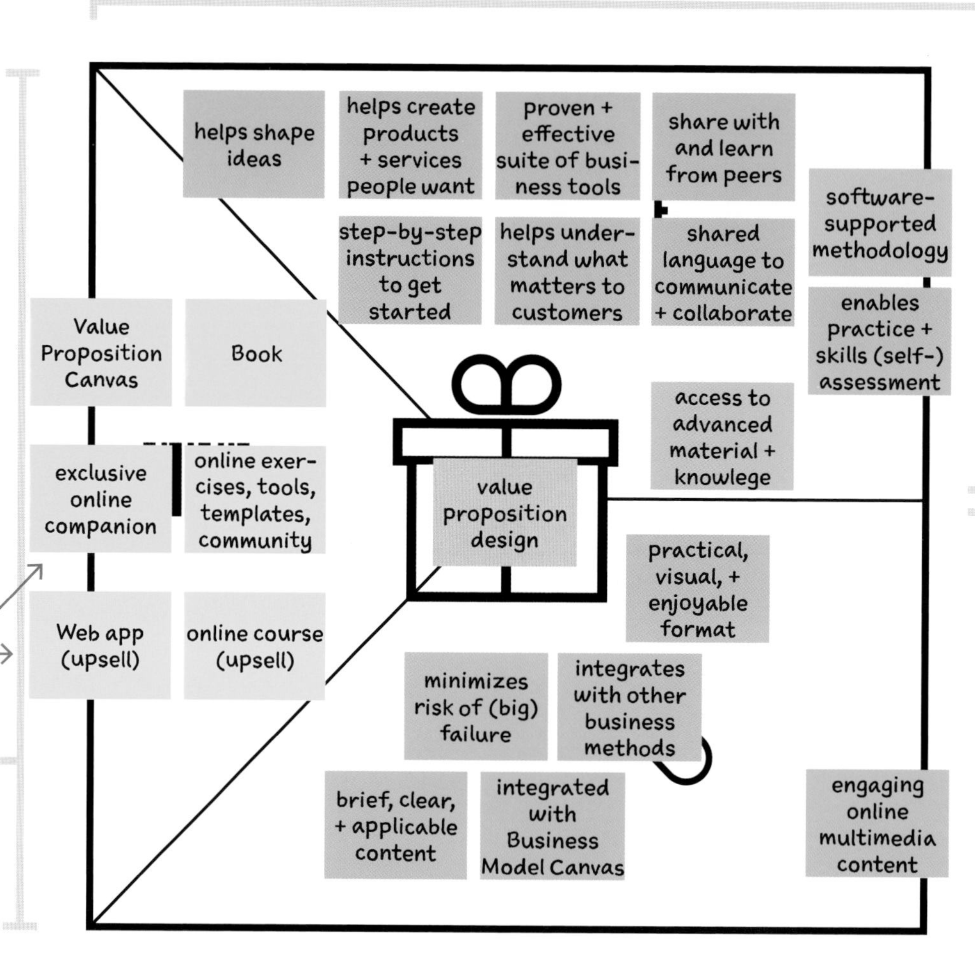

Pain relievers outline how exactly your products and services kill customer pains. Each pain reliever addresses at least one or more pains or gains. Don't add products or services here.

Gain creators highlight how exactly your products and services help customers achieve gains. Each gain creator addresses at least one or more pains or gains. Don't add products or services here.

Formal Map
of how we believe
the products and
services around this
book create value
for customers

Map How Your Products and Services Create Value

OBJECTIVE
Describe explicitly how your products and services create value

OUTCOME
1 page map of value creation

Instructions

Sketch out the value map of one of your existing value propositions. For example, use one that targets the customer segment you profiled in the previous exercise. It's easier to get started with an existing value proposition. However, if you don't have one yet, sketch out how you intend to create value with a new idea. We will cover the creation of new value propositions more specifically later on in this book.

For now:

1. Grab the customer profile you previously completed.
2. Download the value map.
3. Grab a set of small sticky notes.
4. Map out how you create value for your customers.

Download the Value Map pdf

The Value Map

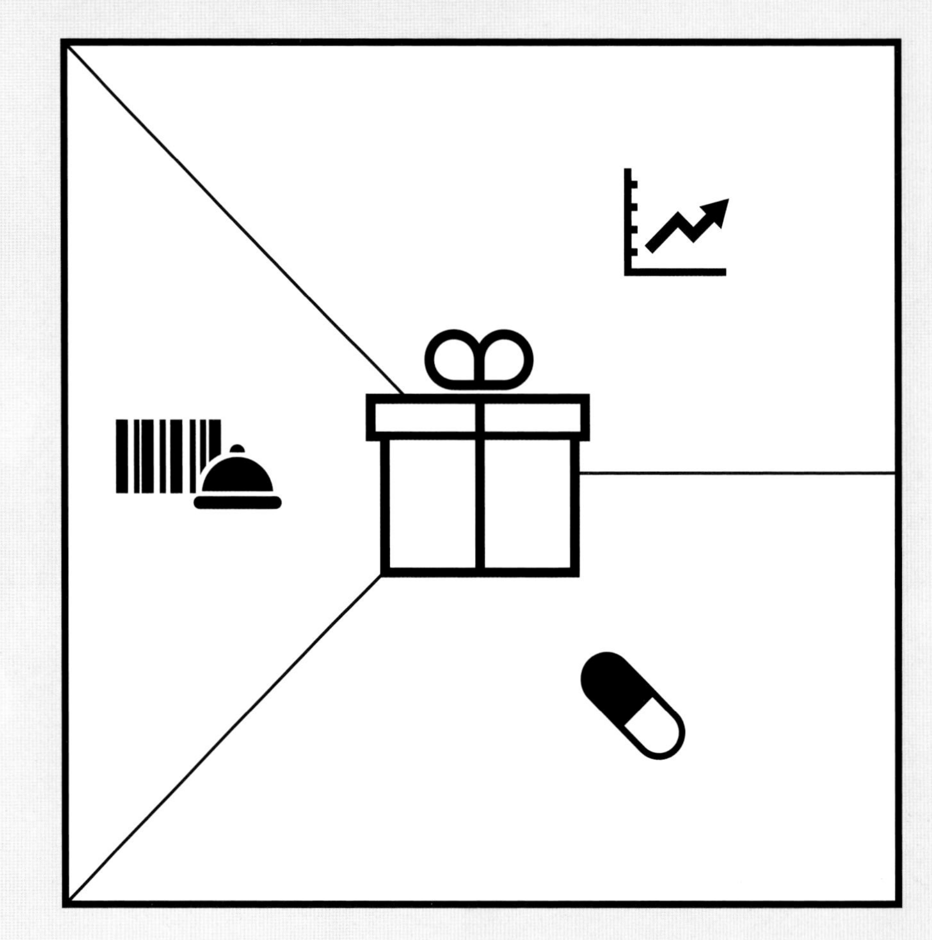

Strategyzer

Copyright Business Model Foundry AG
The makers of Business Model Generation and Strategyzer

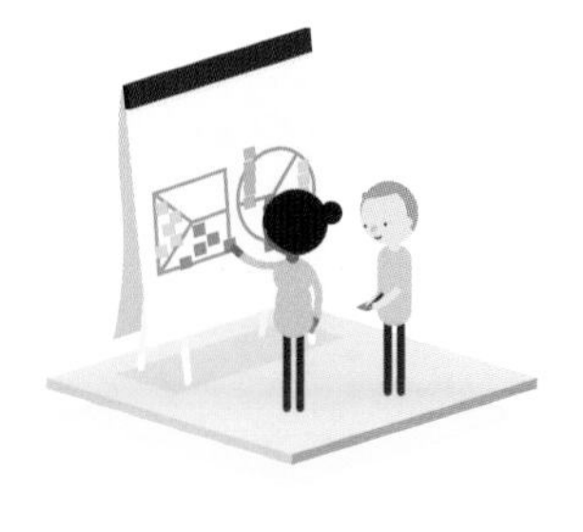

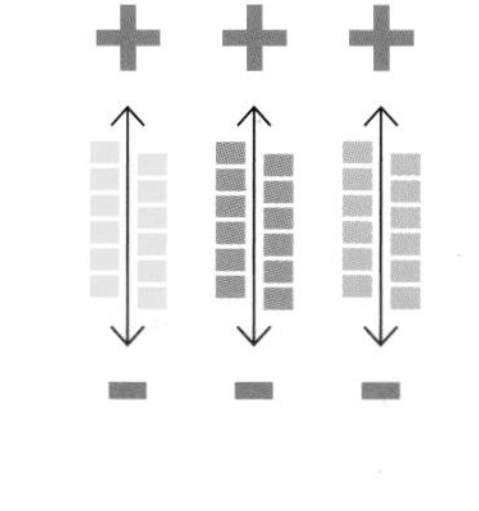

1

List products and services.
List all the products and services of your existing value proposition.

2

Outline pain relievers.
Outline how your products and services currently help customers alleviate pains by eliminating undesired outcomes, obstacles, or risks. Use one sticky note per pain reliever.

3

Outline gain creators.
Explain how your products and services currently create expected or desired outcomes and benefits for customers. Use one sticky note per gain creator.

4

Rank by order of importance.
Rank products and services, pain relievers, and gain creators according to how essential they are to customers.

Do this exercise online

Pain relievers vs. Gain creators

Pain relievers and gain creators both create value for the customer in different ways. The difference is that the former specifically addresses pains in the customer profile, while the latter specifically addresses gains. It is okay if either of them addresses pains and gains at the same time. The main goal of these two areas is to make the customer value creation of your products and services explicit.

?

What is the difference with the pains and gains in the customer profile?

Pain relievers and gain creators are distinctly different from pains and gains. You have control over the former, whereas you don't have control over the latter. You decide (i.e., design) how you intend to create value by addressing specific jobs, pains, and gains. You don't decide over which jobs, pains, and gains the customer has. And no value proposition addresses all of a customer's jobs, pains, and gains. The best ones address those that matter most to customers and do so extremely well.

Best Practices for Mapping Value Creation

✗ Common Mistakes

List all your products and services rather than just those targeted at a specific segment.

Add products and services to the pain reliever and gain creator fields.

Offer pain relievers and gain creators that have nothing to do with the pains and gains in the customer profile.

Make the unrealistic attempt to address all customer pains and gains.

✓ Best Practices

Products and services create value only in relationship to a specific customer segment. List only the bundle of products and services that jointly form a value proposition for a specific customer segment.

Pain relievers and gain creators are explanations or characteristics that make the value creation of your products and services explicit. Examples include "helps save time" and "well-designed."

Remember that products and services don't create value in absolute terms. It is always relative to customers' jobs, pains, and gains.

Realize that great value propositions are about making choices regarding which jobs, pains, and gains to address and which to forgo. No value proposition addresses all of them. If your value map indicates so, it's probably because you're not honest about all the jobs, pains, and gains that should be in your customer profile.

1.3

Fit

Fit

You achieve fit when customers get excited about your value proposition, which happens when you address important jobs, alleviate extreme pains, and create essential gains that customers care about. As we will explain throughout this book, fit is hard to find and maintain. Striving for fit is the essence of value proposition design.

Customers expect and desire a lot from products and services, yet they also know they can't have it all. Focus on those gains that matter most to customers and make a difference.

Customers have a lot of pains. No organization can reasonably address all of them. Focus on those headaches that matter most and are insufficiently addressed.

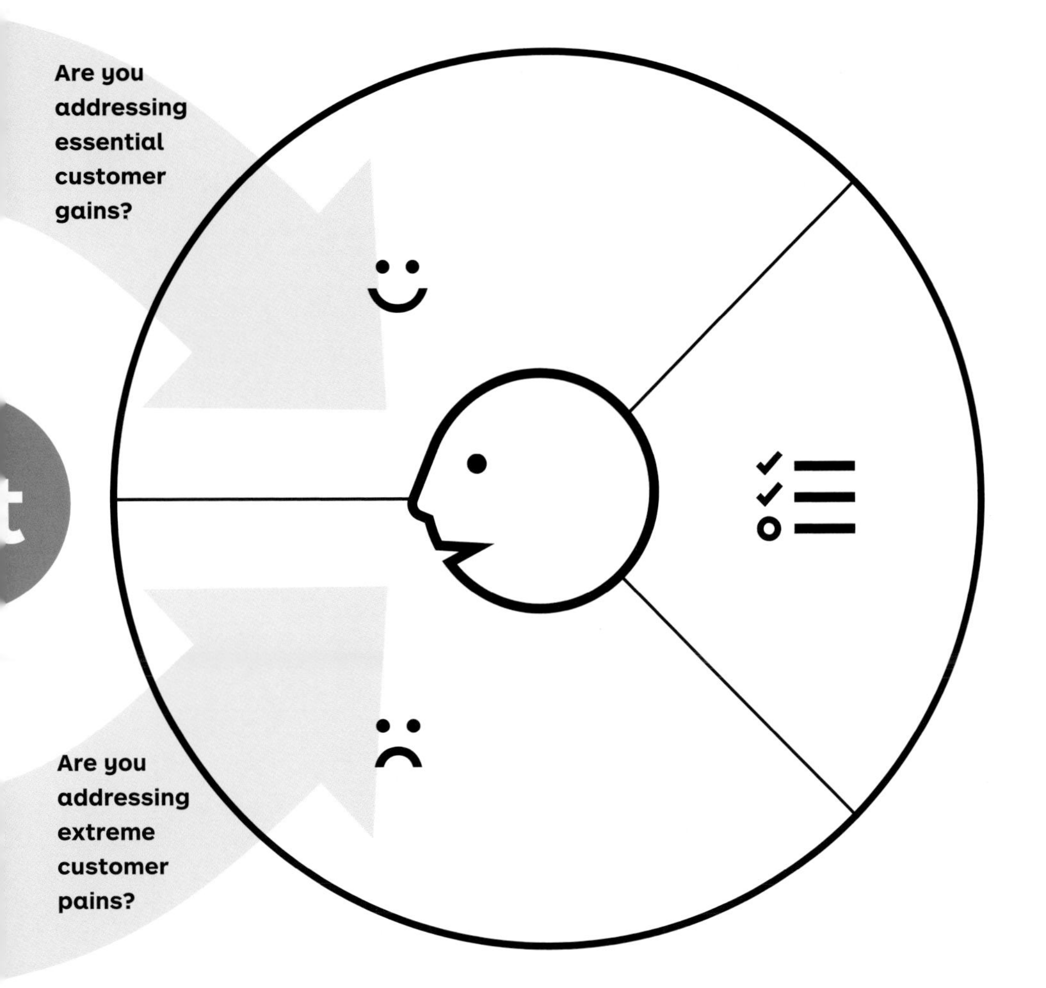

Your customers are the judge, jury, and executioner of your value proposition. They will be merciless if you don't find fit!

Fit?

When we designed the value proposition for this book, we strived to address some of the most important jobs, pains, and gains that potential customers have and that are insufficiently addressed by current business book formats.

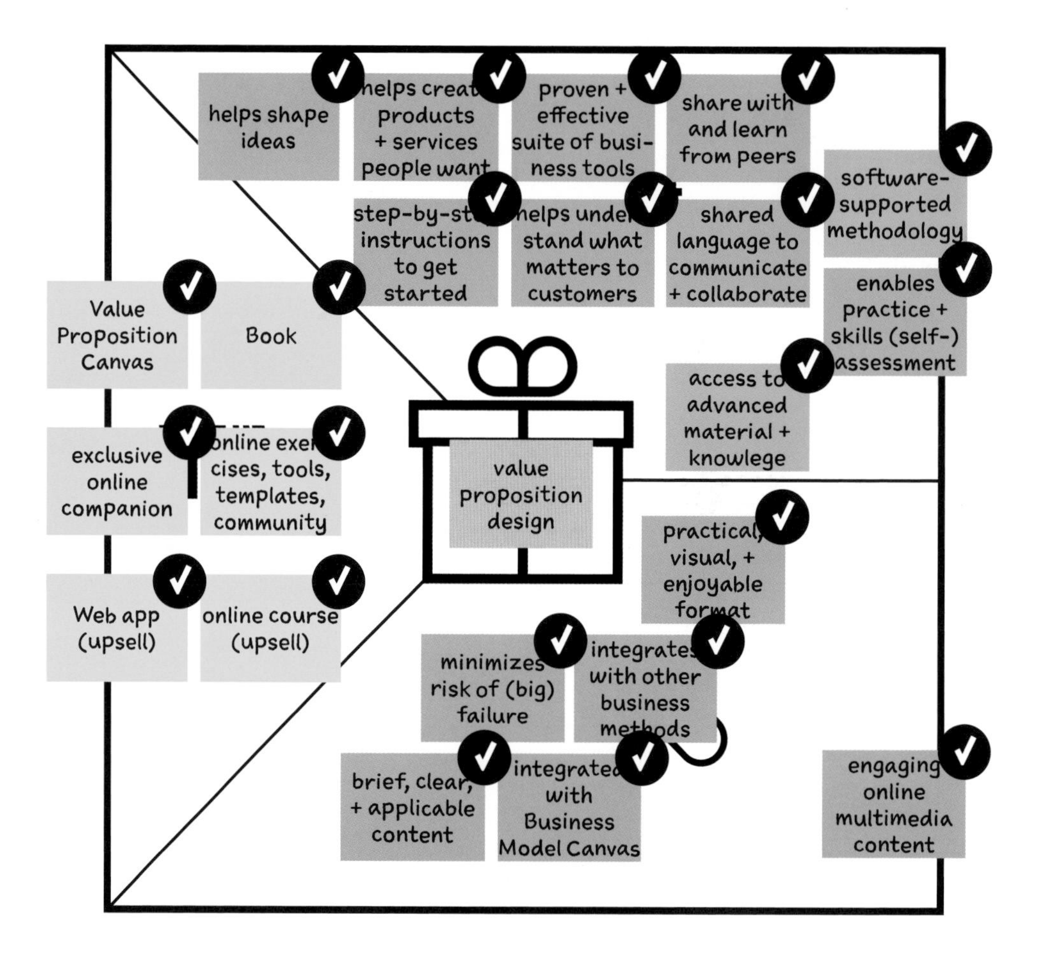

Check marks signify that products and services relieve pains or create gains and directly address one of the customers' jobs, pains, or gains.

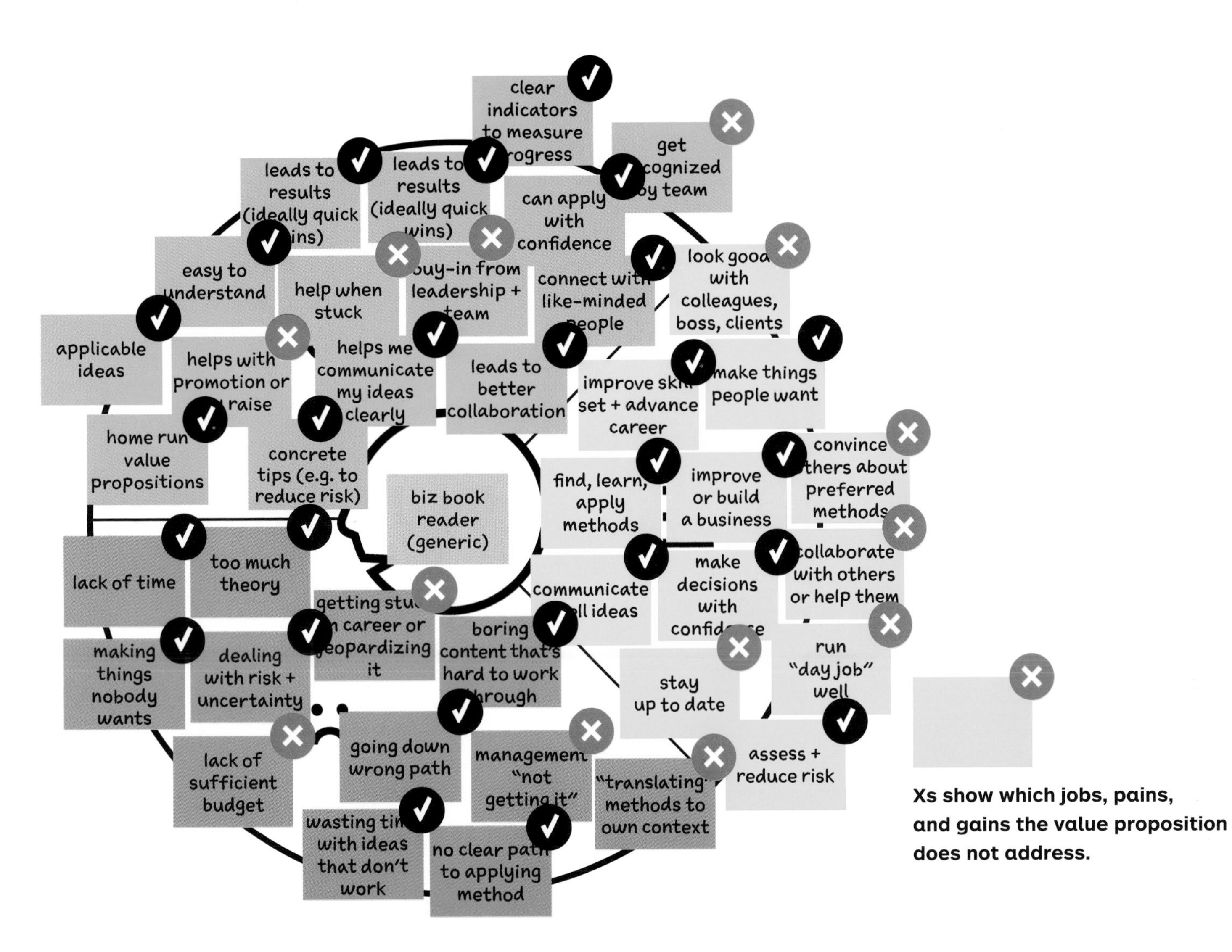

Xs show which jobs, pains, and gains the value proposition does not address.

Check Your Fit

OBJECTIVE	OUTCOME
Verify if you are addressing what matters to customers	Connection between your products and services and customer jobs, pains, and gains

Do this exercise online

1

Instructions

Bring in the Value Proposition Map and Customer Segment Profile you completed earlier. Go through pain relievers and gain creators one by one, and check to see whether they fit a customer job, pain, or gain. Put a check mark on each one that does.

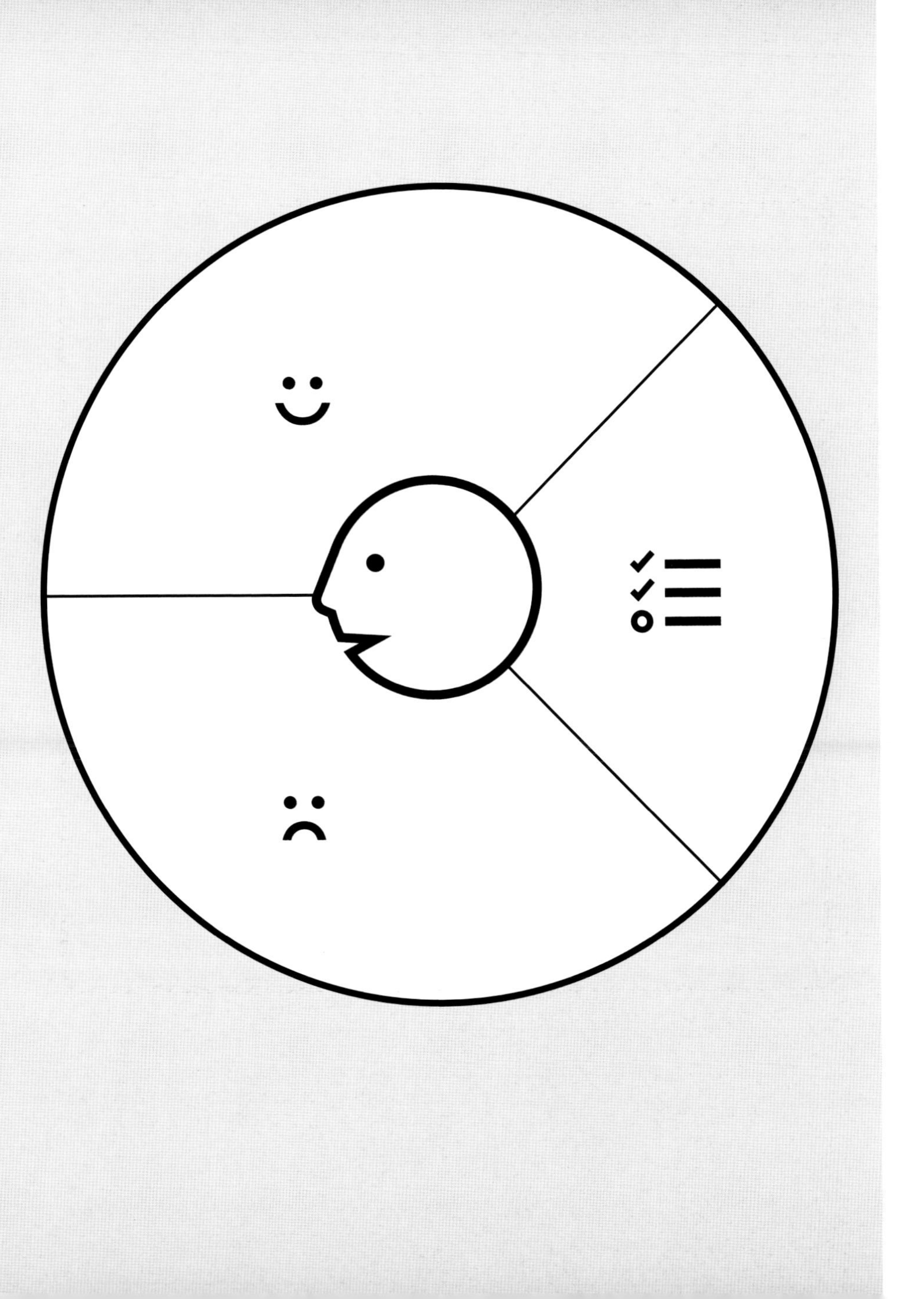

2

Outcome

If a pain reliever or gain creator doesn't fit anything, it may not be creating customer value. Don't worry if you haven't checked all pains/gains—you can't satisfy them all. Ask yourself, how well does your value proposition really fit your customer?

Download the Value Proposition Canvas pdf

Three Kinds of Fit

Searching for Fit is the process of designing value propositions around products and services that meet jobs, pains, and gains that customers really care about. Fit between what a company offers and what customers want is the number one requirement of a successful value proposition.

Fit happens in three stages. The first occurs when you identify relevant customer jobs, pains, and gains you believe you can address with your value proposition. The second occurs when customers positively react to your value proposition and it gets traction in the market. The start-up movement calls these problem-solution fit and product-market fit, respectively. The third occurs when you find a business model that is scalable and profitable.

Get "Fit" poster

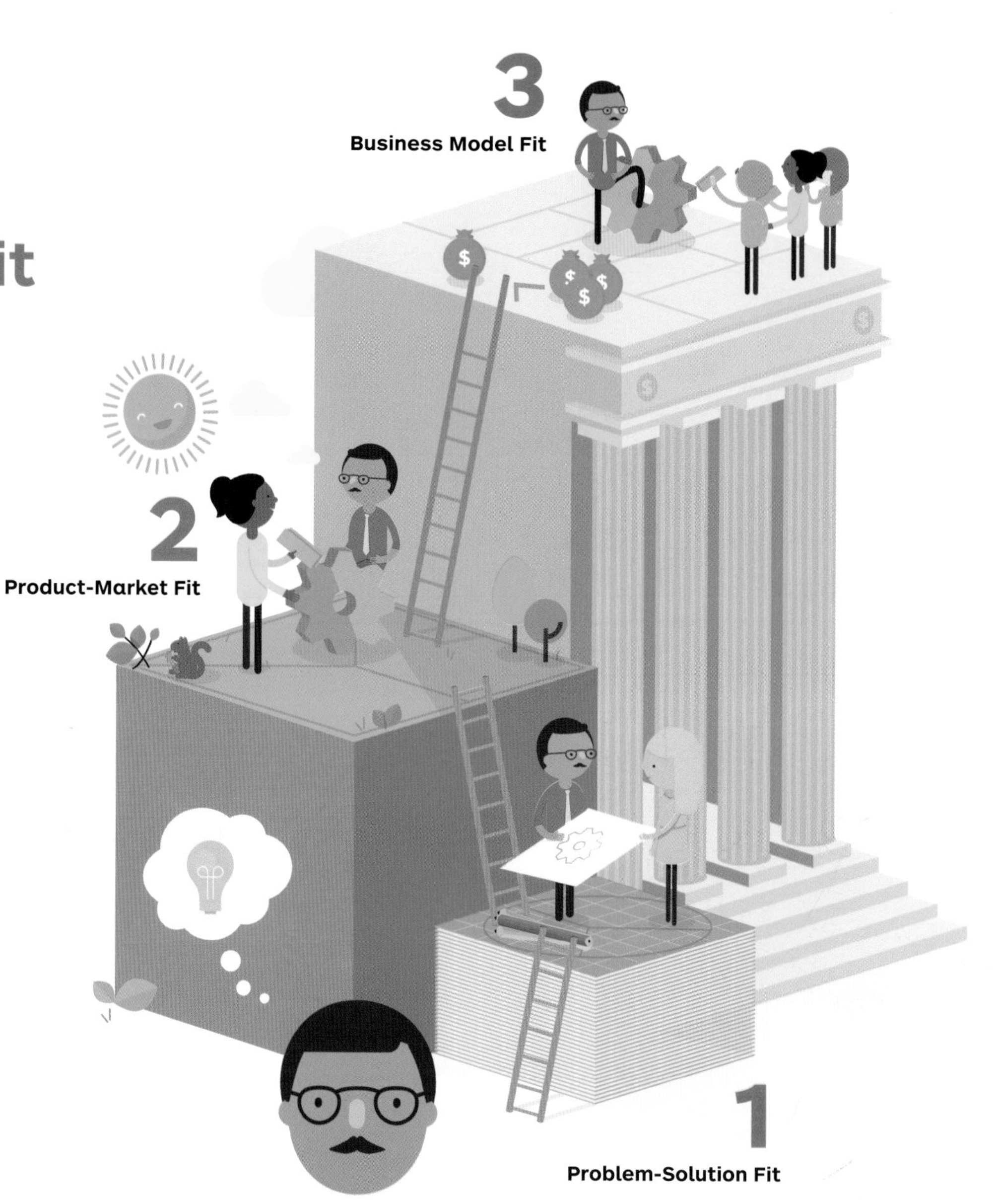

On Paper →

1. Problem-Solution Fit

Problem-solution fit takes place when you

- Have evidence that customers care about certain jobs, pains, and gains.
- Designed a value proposition that addresses those jobs, pains, and gains.

At this stage you don't yet have evidence that customers actually care about your value proposition.

This is when you strive to identify the jobs, pains, and gains that are most relevant to customers and design value propositions accordingly. You prototype multiple alternative value propositions to come up with the ones that produce the best fit. The fit you achieve is not yet proven and exists mainly on paper. Your next steps are to provide evidence that customers care about your value proposition or start over with designing a new one.

In the Market →

2. Product-Market Fit

Product-market fit takes place when you

- Have evidence that your products and services, pain relievers, and gain creators are actually creating customer value and getting traction in the market.

During this second phase, you strive to validate or invalidate the assumptions underlying your value proposition. You will inevitably learn that many of your early ideas simply don't create customer value (i.e., customers don't care) and will have to design new value propositions. Finding this second type of fit is a long and iterative process; it doesn't happen overnight.

In the Bank

3. Business Model Fit

Business model fit takes place when you

- Have evidence that your value proposition can be embedded in a profitable and scalable business model.

A great value proposition without a great business model may mean suboptimal financial success or even lead to failure. No value proposition—however great—can survive without a sound business model.

The search for business model fit entails a laborious back and forth between designing a value proposition that creates value for customers and a business model that creates value for your organization. You don't have business model fit until you can generate more revenues with your value proposition than you incur costs to create and deliver it (or "them" in the case of platform models with more than one interdependent value propositions).

Customer Profiles in B2B

Value propositions in business-to-business (B2B) transactions typically involve several stakeholders in the search, evaluation, purchase, and use of a product or service. Each one has a different profile with different jobs, pains, and gains. Stakeholders can tilt the purchasing decision in one direction or another. Identify the most important ones and sketch out a Value Proposition Canvas for each one of them.

Profiles vary according to the sector and size of organization, but they typically include the following roles:

+Unbundled

Value propositions to stakeholders *within* the business

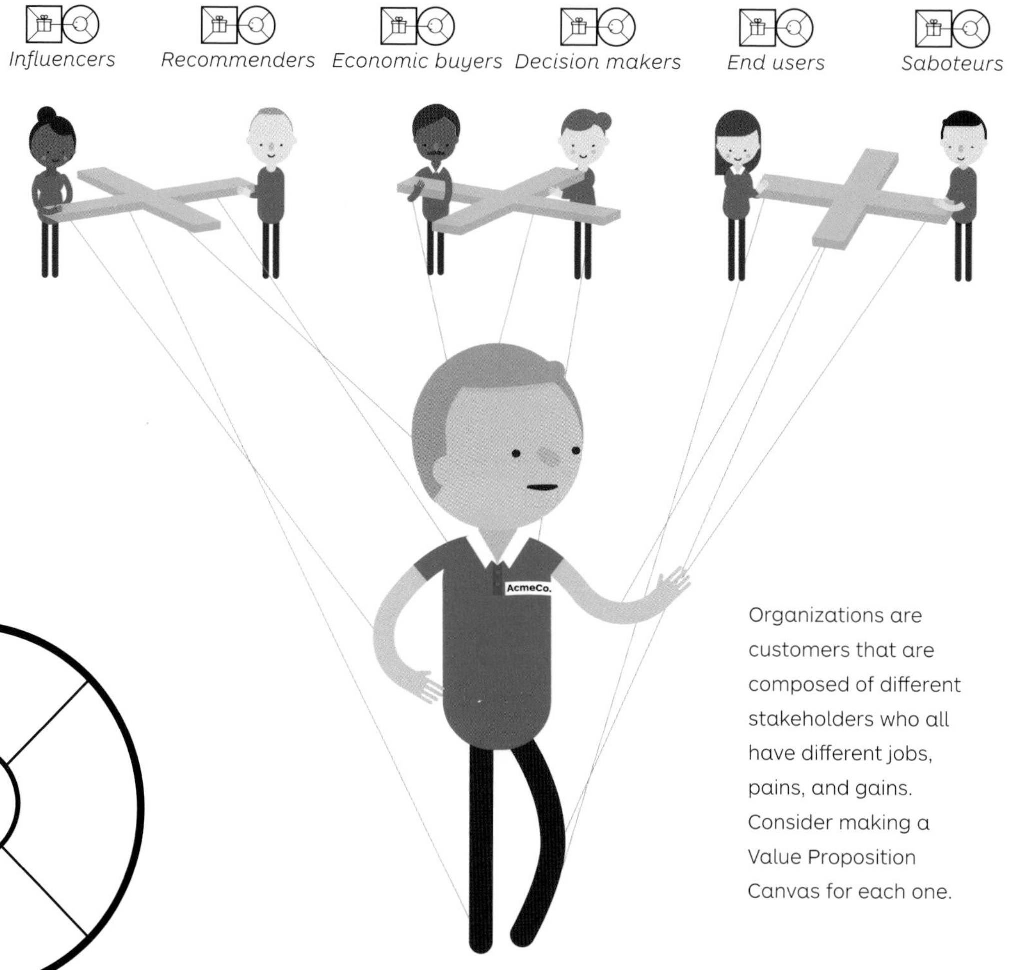

Organizations are customers that are composed of different stakeholders who all have different jobs, pains, and gains. Consider making a Value Proposition Canvas for each one.

Aggregated

Value Proposition

Business Segment

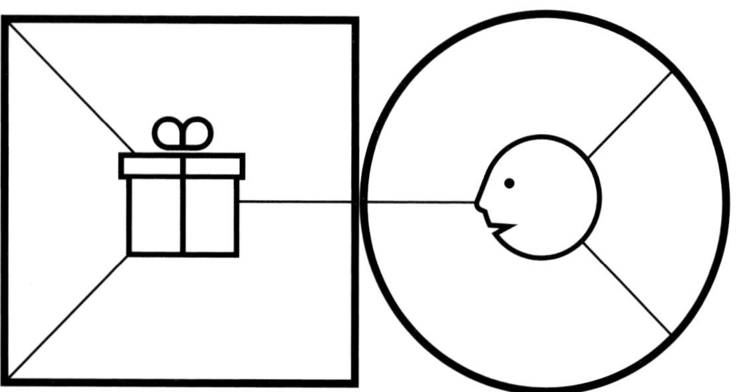

Adapted from Steve Blank, The Four Steps to the Epiphany, 2006.

Influencers
Individuals or groups whose opinions might count and whom the decision maker might listen to, even in an informal way.

Recommenders
The people carrying out the search and evaluation process and who make a formal recommendation for or against a purchase.

Economic Buyers
The individual or group who controls the budget and who makes the actual purchase. Their concerns are typically about financial performance and budgetary efficiency.

In some cases, the economic buyer may sit outside an organization, such as a government paying for the basic medical supply in nursing homes for elderly citizens.

Decision Makers
The person or group ultimately responsible for the choice of a product/service and for ordering the purchase decision. Decision makers usually have ultimate authority over the budget.

End Users
The ultimate beneficiaries of a product or service. For a business customer, end users can either be within their own organization (a manufacturer buying software for its designers), or they can be external customers (a device manufacturer buying chips for the smartphones it sells to consumers). End users may be passive or active, depending on how much say they have in the decision and purchase process.

Saboteurs
The people and groups who can obstruct or derail the process of searching, evaluating, and purchasing a product or a service.

Decision makers typically sit inside the customer's organization, whereas Influencers, recommenders, economic buyers, end users, and saboteurs can sit inside or outside the organization.

Unbundling the Family

Value propositions to the consumer may also involve several stakeholders in the search, evaluation, purchase, and use of a product or service. For example, consider a family that intends to buy a game console. In this situation, there is also a difference between the economic buyer, the influencer, the decision maker, the users, and the saboteurs. It therefore makes sense to sketch out a different Value Proposition Canvas for each stakeholder.

Multiple Fits

Some business models work only with a combination of several value propositions and customer segments. In these situations, you require fit between each value proposition and its respective customer segment for the business model to work.

Two common illustrations of multiple fits are *intermediary* and *platform* business models.

Intermediary

When a business sells a product or service through an intermediary, it effectively needs to cater to two customers: the end customer and the intermediary itself. Without a clear value proposition to the intermediary, the offer might not reach the end customer at all, or at least not with the same impact.

Chinese firm Haier sells home appliances and electronics to households globally. It does this largely through retailers such as Carrefour, Walmart, and others. To be successful, Haier needs to craft an appealing value proposition both to households (the end customer) and to intermediary distributors.

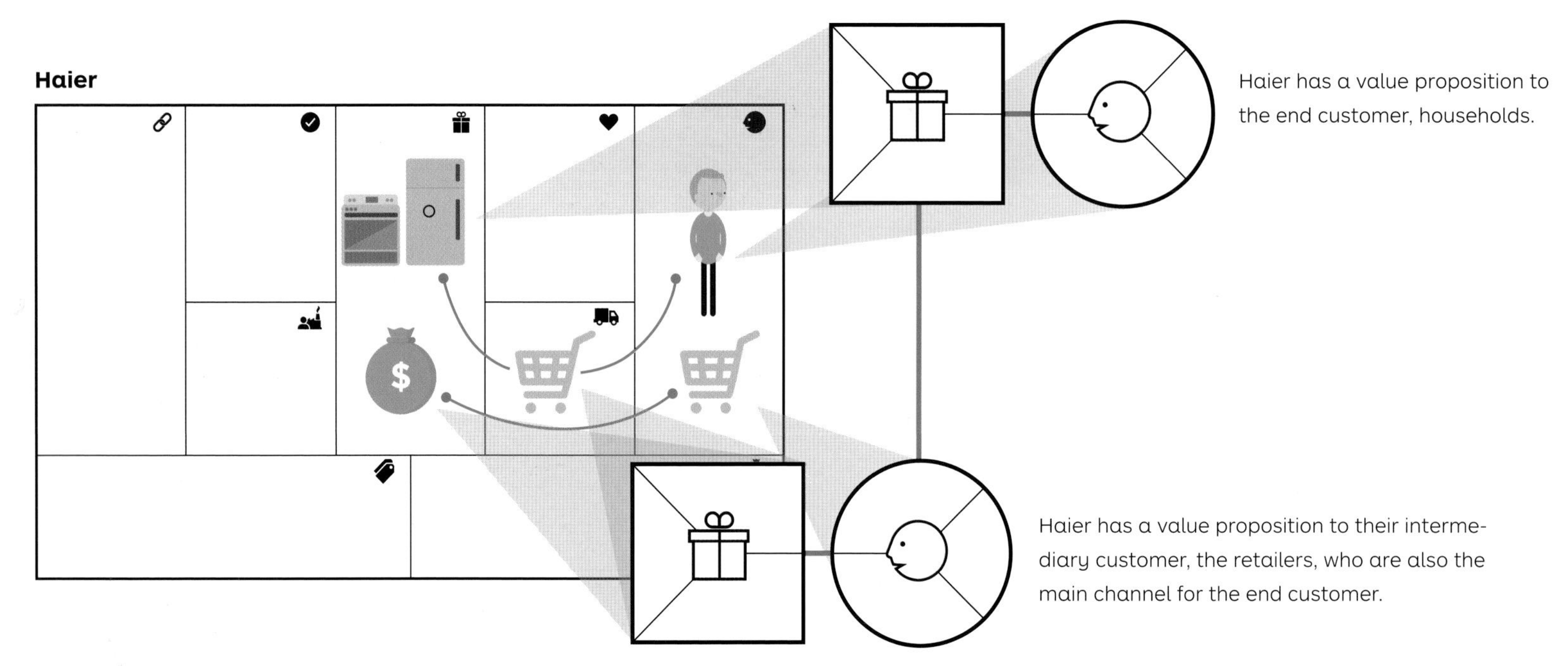

Haier has a value proposition to the end customer, households.

Haier has a value proposition to their intermediary customer, the retailers, who are also the main channel for the end customer.

Platforms

Platforms function only when two or more actors interact and draw value within the same interdependent business model. Platforms are called double-sided when there are two such actors and multisided when there are more than two. A platform exists only when all sides are present in the model.

Airbnb is an example of a double-sided platform. It is a website that connects local residents with extra space to rent out and travelers looking for alternatives to hotels as a place to stay. In such a case, the business model needs to hold two value propositions, one for local residents (called hosts) and one for travelers.

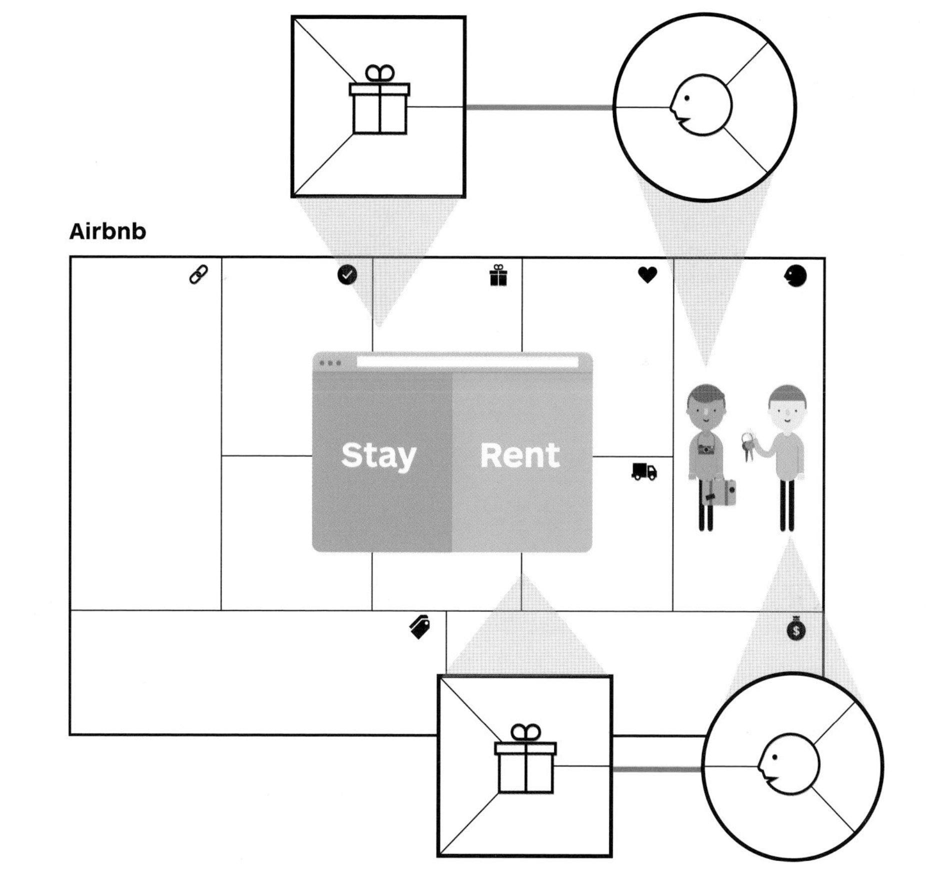

Going to the Movies

Let us walk through the concepts of the Value Proposition Canvas with another simple example. Imagine the owner of a movie theater chain wants to design new value propositions for his customers.

He could start with the value proposition's features and get excited about the latest generation of big screens, state-of-the-art display technologies, tasty snacks, social happenings, urban experiences, and so on. But, of course, those only really matter if customers care about them. So he sets out to better understand what his customers truly want.

Traditionally he'd sketch out psychodemographic profiles of his customer segments. But this time he decides to complement this type of segmentation with customer profiles that highlight a customer's jobs, pains, and gains.

What drives the moviegoer?

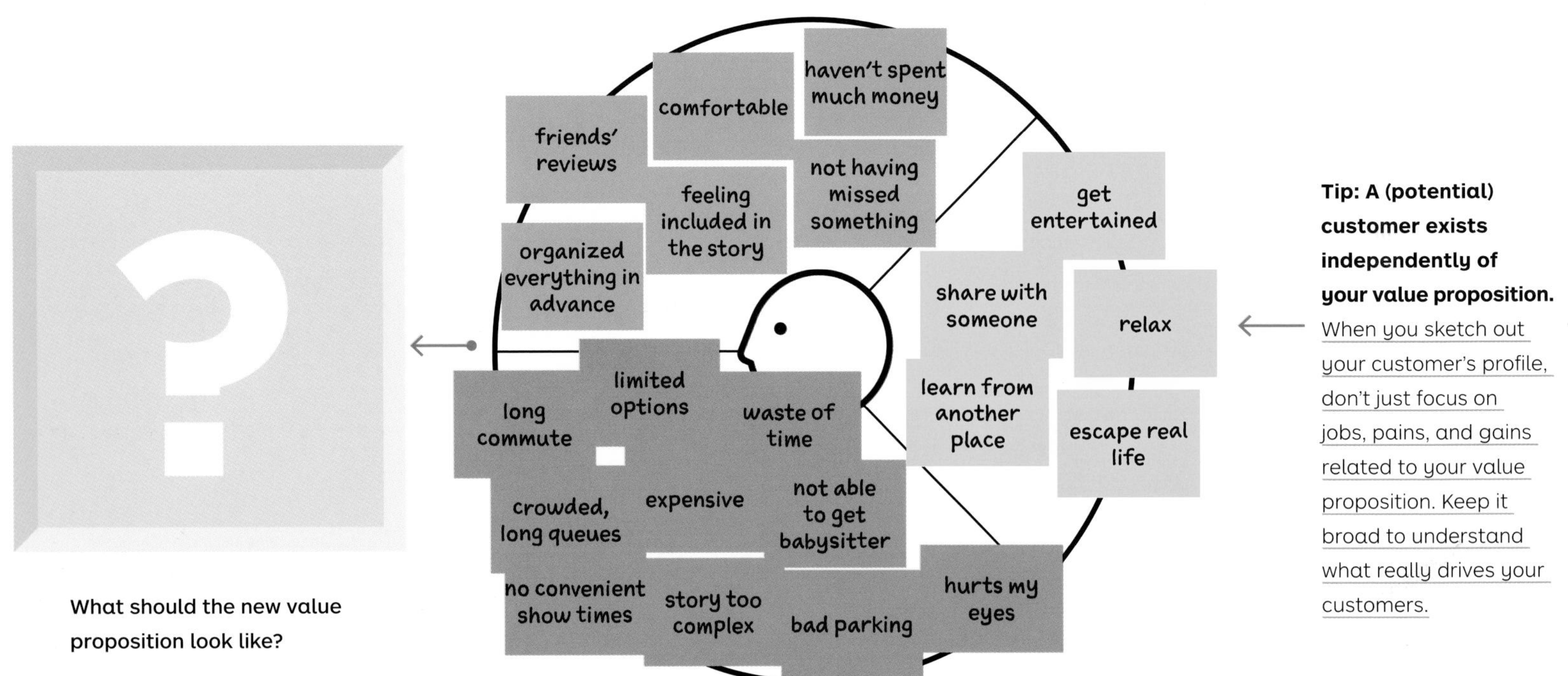

What should the new value proposition look like?

Tip: A (potential) customer exists independently of your value proposition. When you sketch out your customer's profile, don't just focus on jobs, pains, and gains related to your value proposition. Keep it broad to understand what really drives your customers.

A Movie Theater's Business Model

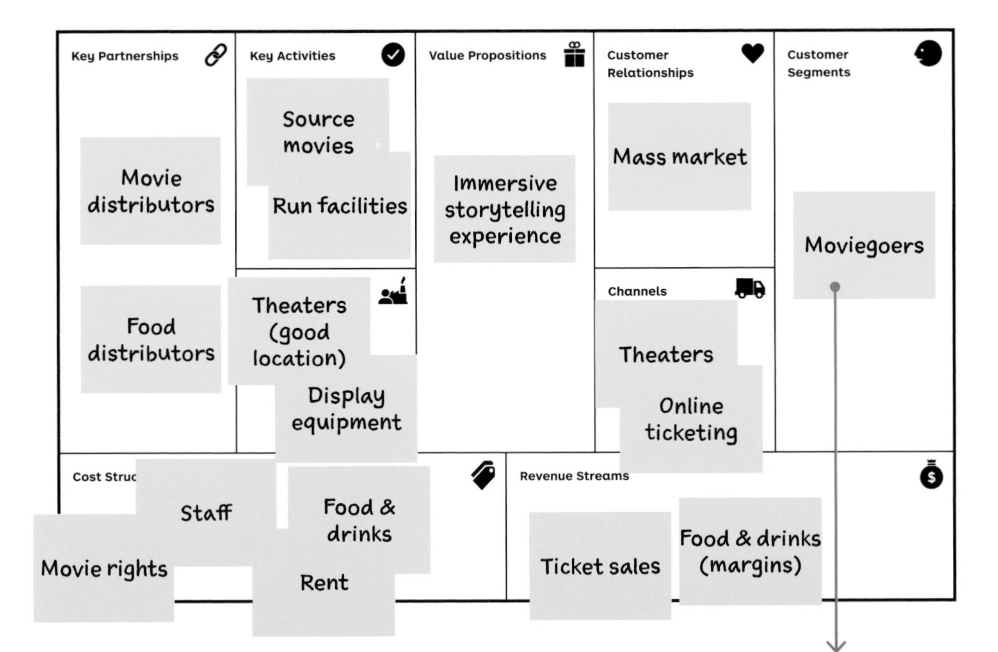

The new approach: *focusing on the jobs, pains, and gains that drive customers*

By sketching out a customer profile, you aim to uncover what really drives people, rather than just describing their socioeconomic characteristics. You investigate what they're trying to achieve, their underlying motives, their objectives, and what's holding them back. Doing so will broaden your horizon and likely uncover new or better opportunities to satisfy customers.

The traditional approach: *psychodemographic profiles*

Traditional psychodemographic profiles group consumers into categories that have the same socioeconomic characteristics.

JANE MOVIEGOER
20-30 years old
Upper middle class
Earns $100K/year
Married, 2 children

Movie Behavior:

- Prefers action movies
- Likes popcorn and soda
- Does not like waiting in line
- Buys tickets online
- Goes once a month

Same Customer, Different Contexts

Priorities change depending on a customer's context. Taking this context into account before you think of a value proposition for that customer is crucial.

With the jobs-to-be-done approach, you uncover the motivations of different customer segments. Yet, depending on the context, some jobs will become more important or matter less than others.

In fact, the context in which a person finds himself or herself often changes the nature of the jobs that the person aims to accomplish.

For example, the clientele of a restaurant is likely to use very different criteria to evaluate their dining experience at lunch versus at dinner. Likewise a mobile phone user will have different job requirements when using the phone in a car, in a meeting, or at home. Hence, the features of your value proposition will be different depending on which context(s) you are focusing.

In our example, the context in which our moviegoer finds herself will influence which jobs matter more or less to her.

Add contextual elements to your customer profiles if necessary. They might serve as constraints for designing value propositions later on.

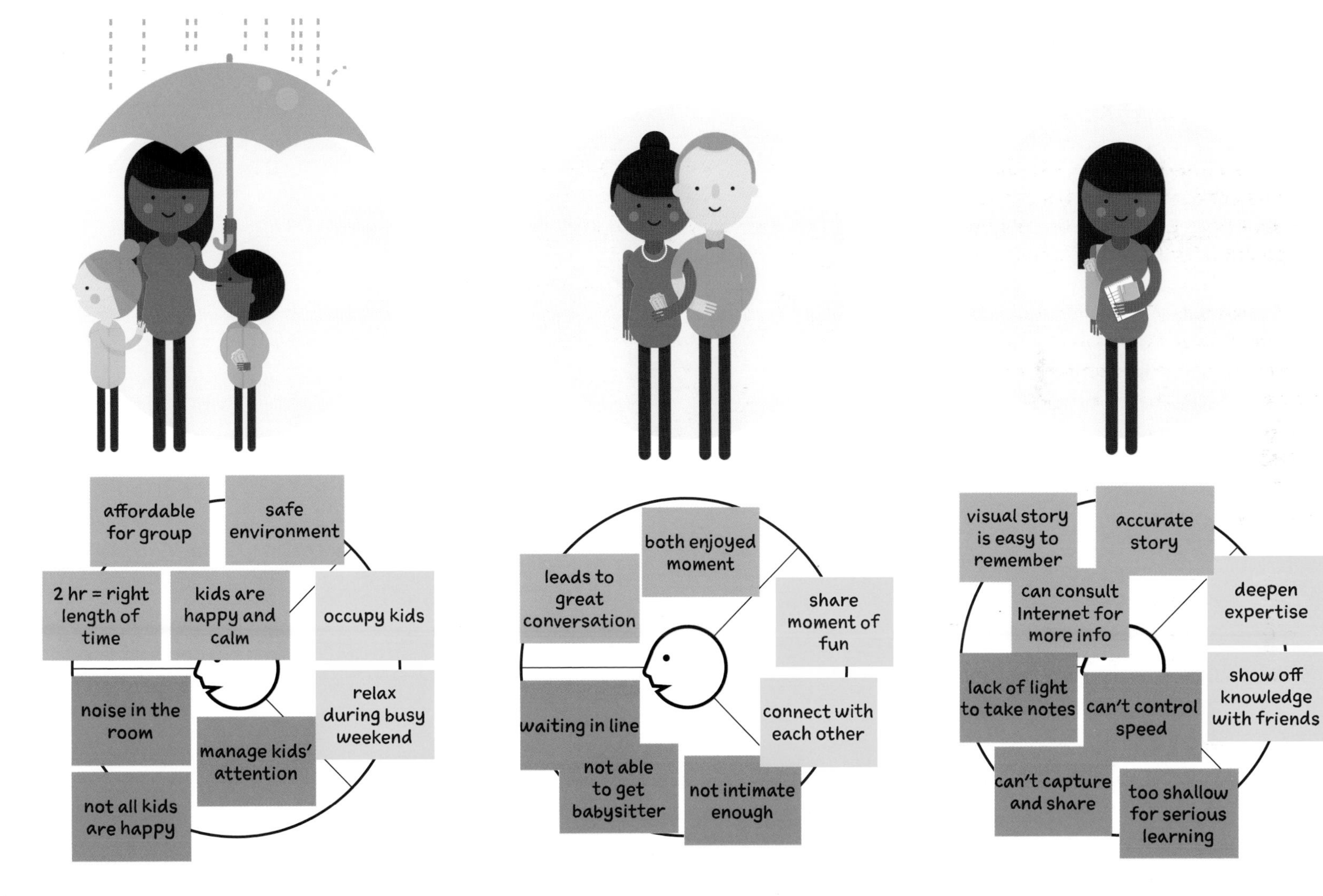

The kids' afternoon off

When? Wednesday afternoon

Where? Leaving from home

With whom? Kids and maybe their friends

Constraints? After school, before dinner time

Date night

When? Saturday evening

Where? Leaving from home

With whom? Partner

Constraints? Kids taken care of (if parents)

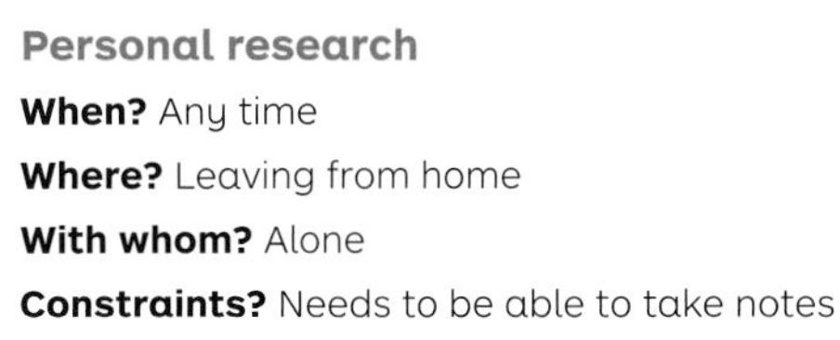

Personal research

When? Any time

Where? Leaving from home

With whom? Alone

Constraints? Needs to be able to take notes

Same Customer, Different Solutions

In today's hypercompetitive world, customers are surrounded by an ocean of tempting value propositions that all compete for the same limited slots of attention.

Very different value propositions may address similar jobs, pains, and gains. For example, our movie theater chain competes for customer attention not only with other movie theaters but also with a broad range of alternative options: renting a movie at home, going out to dinner, visiting a spa, or maybe even attending an online virtual art exhibit with 3D glasses.

Strive to understand what your customers really care about. Investigate their jobs, pains, and gains beyond what your own value proposition directly addresses in order to imagine totally new or substantially improved ones.

Understand your customers beyond your solution. Unearth the jobs, pains, and gains that matter to them in order to understand how to improve your value proposition or invent new ones.

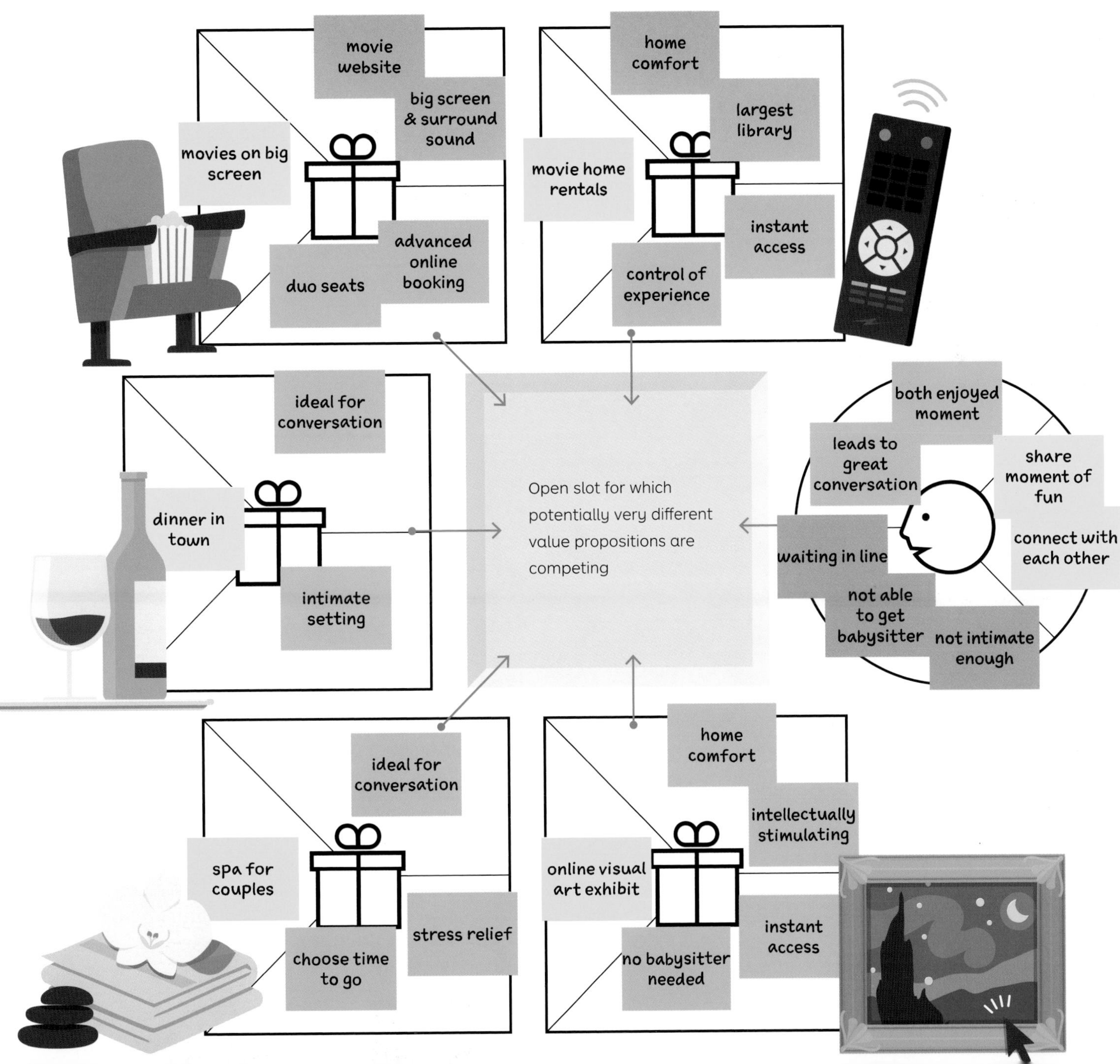
movie website
big screen & surround sound
movies on big screen
advanced online booking
duo seats
home comfort
largest library
movie home rentals
instant access
control of experience
ideal for conversation
dinner in town
intimate setting
Open slot for which potentially very different value propositions are competing
both enjoyed moment
leads to great conversation
share moment of fun
connect with each other
waiting in line
not able to get babysitter
not intimate enough
ideal for conversation
spa for couples
stress relief
choose time to go
home comfort
intellectually stimulating
online visual art exhibit
instant access
no babysitter needed

Lessons Learned

Customer Profile

Use the customer profile to visualize what matters to customers. Specify their jobs, pains, and gains. Communicate the profile across your organization as a one-page actionable document that creates a shared customer understanding. Apply it as a "scoreboard" to track if assumed customer jobs, pains, and gains exist when you talk to real customers.

Value Map

Use the value map to make explicit how you believe your products and services will ease pains and create gains. Communicate the map across your organization as a one-page document that creates a shared understanding of how you intend to create value. Apply it as a "scoreboard" to track if your products actually ease pains and gains when you test them with customers.

Fit

Problem-solution fit: Evidence that customers care about the jobs, pains, and gains you intend to address with your value proposition. Product-market fit: Evidence that customers want your value proposition. Business model fit: Evidence that the business model for your value proposition is scalable and profitable.

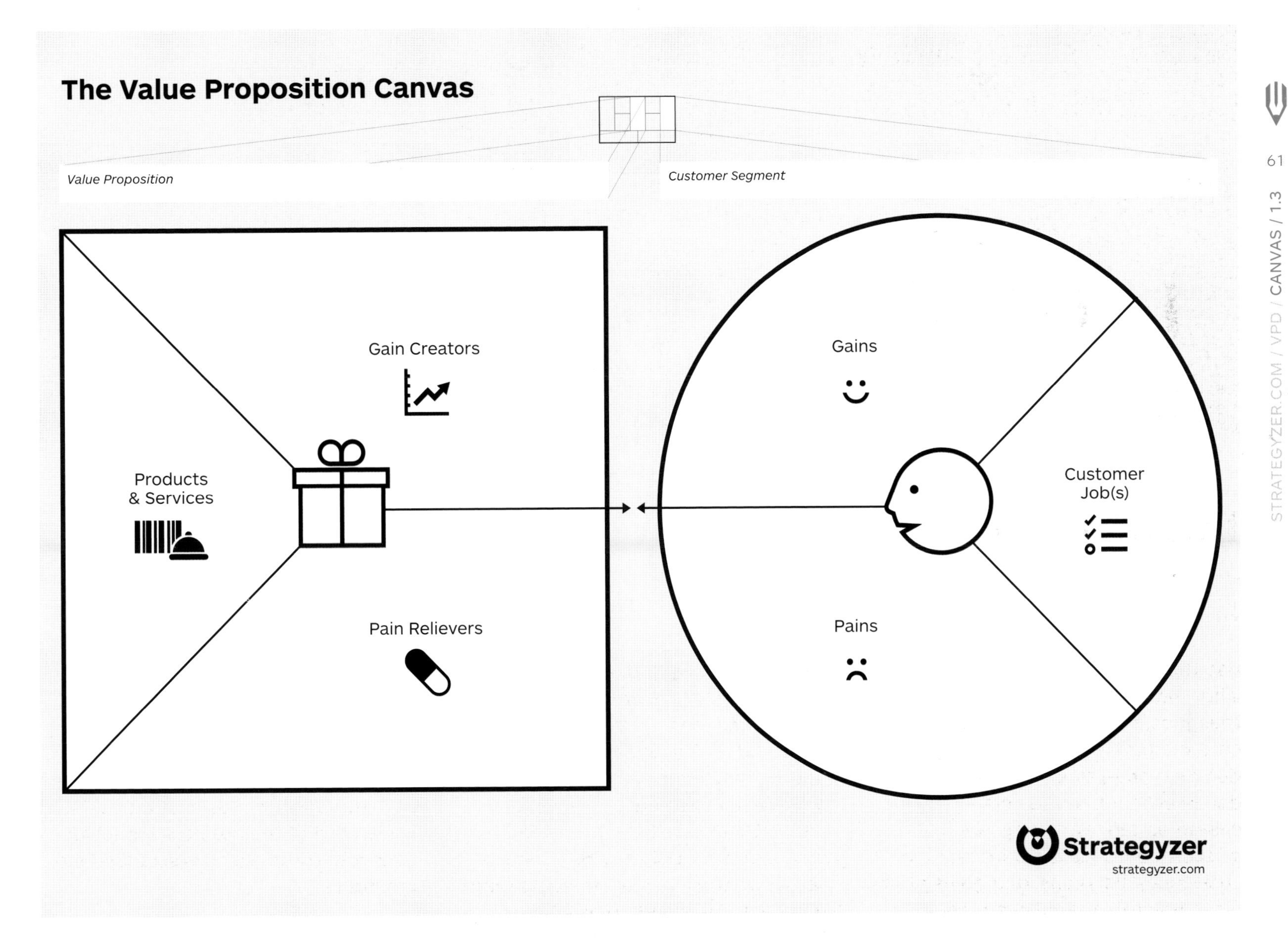

Download the Value Proposition Canvas pdf

Design, Test, Repeat

The search for value propositions that meet customer jobs, pains, and gains is a continuous back and forth between designing prototypes and testing them. The process is iterative rather than sequential. The goal of *Value Proposition Design* is to test ideas as quickly as possible in order to learn, create better designs, and test again.

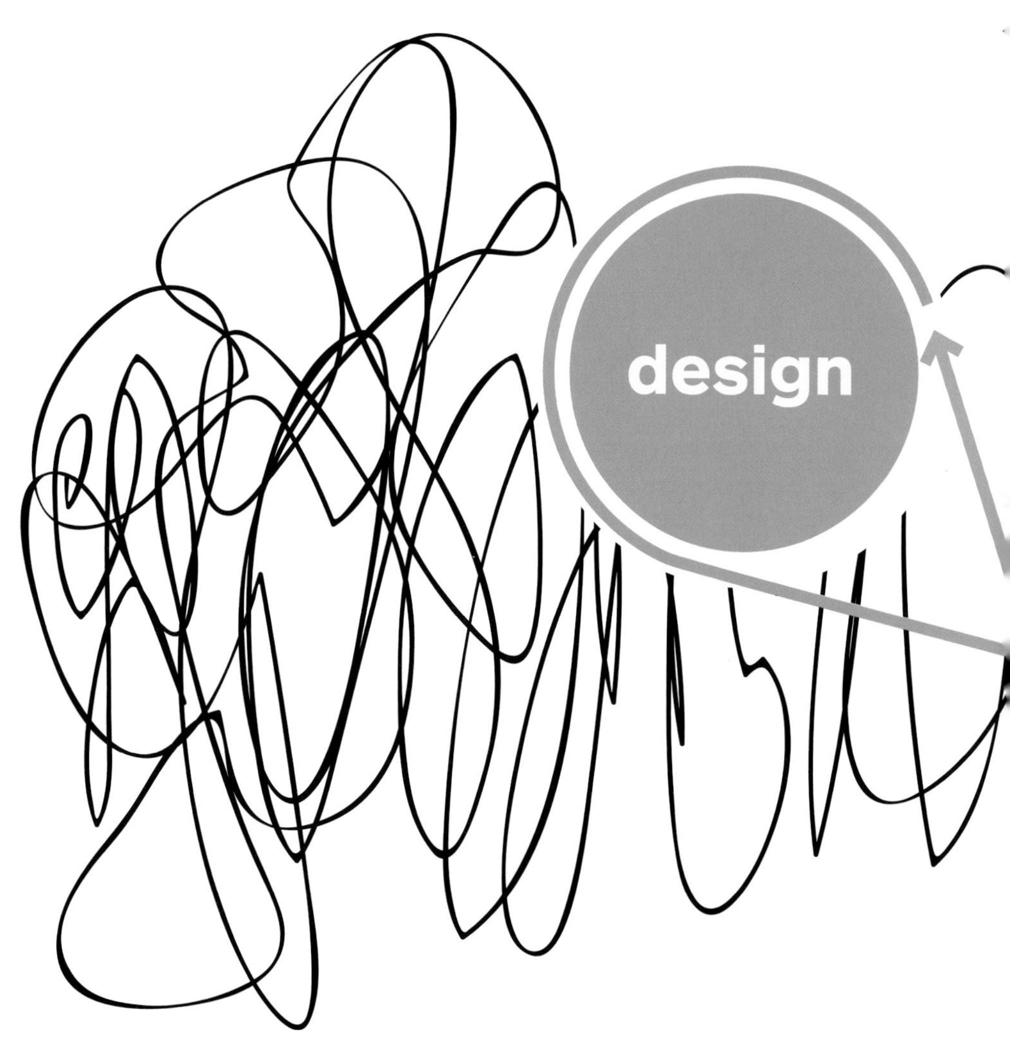

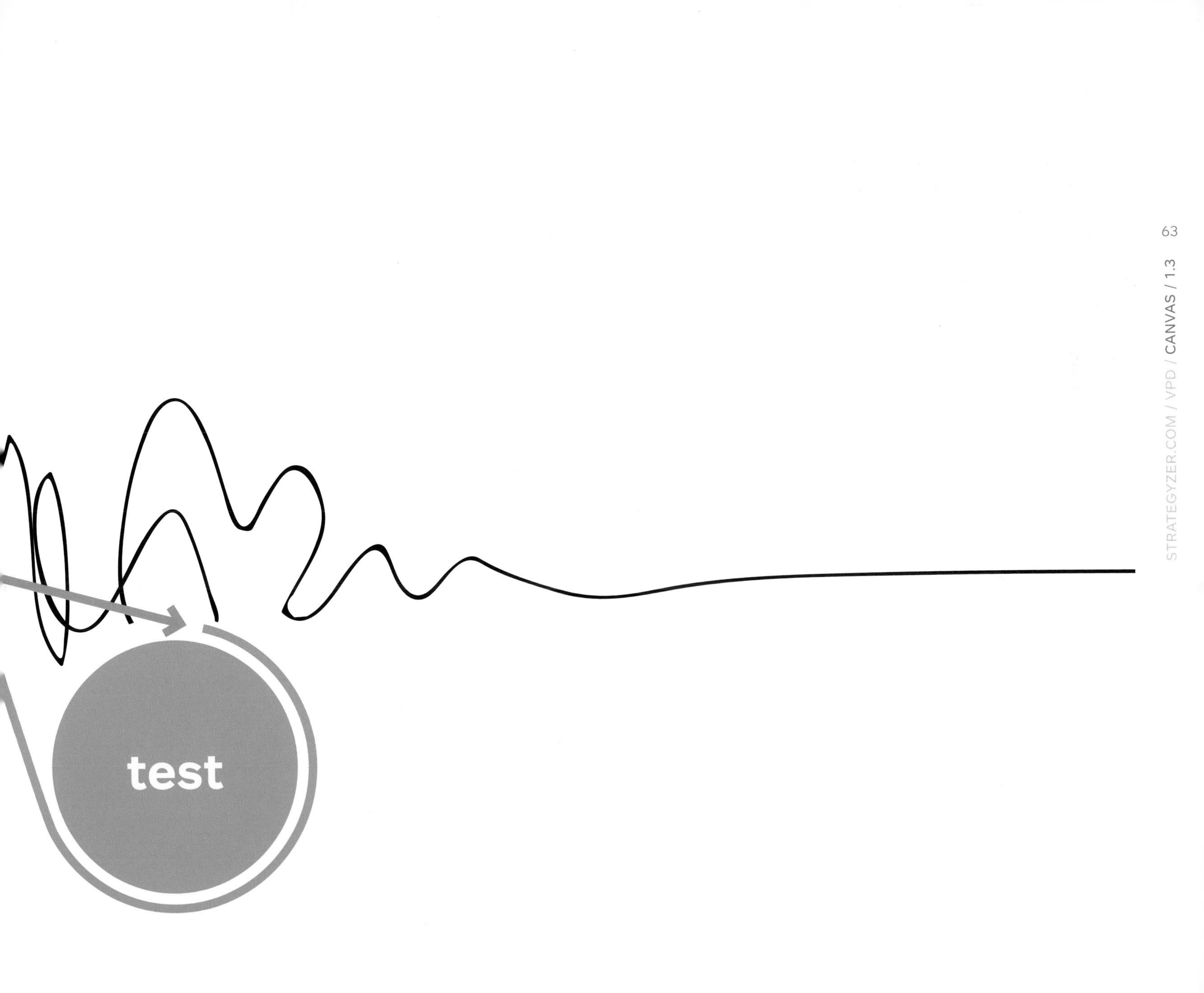
test

des

ign

2

Kick-start value proposition design with Prototyping Possibilities p. 74 for one of your Starting Points p. 86. Shape your value propositions by Understanding Customers p. 104, then select which ones you want to further explore by Making Choices p. 120 and Finding the Right Business Model p. 142. If you are an existing company, discover the particularities of Designing in Established Organizations p. 158.

FROM TESTING
STARTING POINTS
OBSERVATION
Learning Card

Acme
ASSESSMENT
TO TESTING
7 QUESTIONS
FINANCE
PROTOTYPING
BUSINESS MODEL
FAILED

Shaping Your Ideas

Design is the activity of turning your ideas into value proposition prototypes. It is a continuous cycle of prototyping, researching customers, and reshaping your ideas. Design may start with prototyping or with customer discovery. The design activity feeds into the testing activity that we explore in the next chapter (see section 3. Test, ➔ p. 172).

Ideas, Starting Points, and Insights

➔ p. 86

Starting points for new or improved value propositions may come from anywhere. It could be from your customer insights ➔ p. 116, from exploration of prototypes ➔ p. 76, or from many other sources ➔ p. 88. Be sure not to fall in love with your early ideas, because they are certain to transform radically during prototyping ➔ p. 76, customer research ➔ p. 104, and testing ➔ p. 172.

Prototype Possibilities

➔ p. 74

Shape your ideas with quick, cheap, and rough prototypes. Make them tangible with napkin sketches ➔ p. 80, ad-libs ➔ p. 82, and Value Proposition Canvases ➔ p. 84. Don't get attached to a prototype too early. Keep your prototypes light so you can explore possibilities, easily throw them away again, and then find the best ones that survive a rigorous testing process with customers ➔ p. 240.

Understand Customers

➔ p. 104

Inform your ideas and prototypes with early customer research. Plough through available data ➔ p. 108, talk to customers ➔ p. 110, and immerse yourself in their world ➔ p. 114. Don't show customers your value proposition prototypes too early. Use early research to deeply understand your customers' jobs, pains, and gains. Unearth what really matters to them to prototype value propositions that are likely to survive rigorous testing with customers ➔ p. 172.

Prototyping Possibilities
Ideas and Starting Points
Understanding Customers

10 Characteristics of Great Value Propositions

Stop for an instant and reflect on the characteristics of great value propositions before reading about how to design them in this chapter. We offer 10 characteristics to get you started. Don't hesitate to add your own. Great Value Propositions...

Get "10 Characteristics of Great Value Propositions" poster

1
Are embedded in great business models

2
Focus on the jobs, pains, and gains that matter most to customers

3
Focus on unsatisfied jobs, unresolved pains, and unrealized gains

4
Target few jobs, pains, and gains, but do so extremely well

5
Go beyond functional jobs and address emotional and social jobs

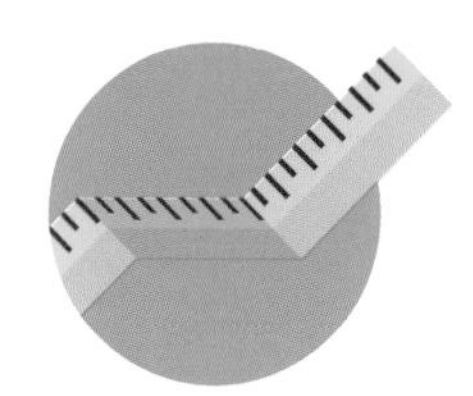

6
Align with how customers measure success

7
Focus on jobs, pains, and gains that a lot of people have or that some will pay a lot of money for

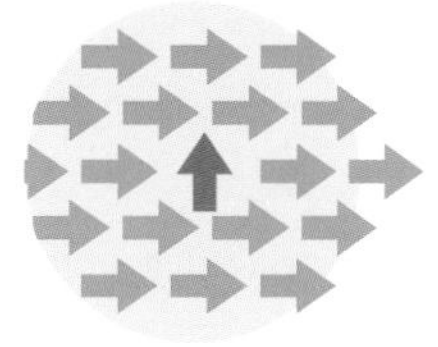

8
Differentiate from competition on jobs, pains, and gains that customers care about

9
Outperform competition substantially on at least one dimension

10
Are difficult to copy

2.1

Prototyping Possibilities

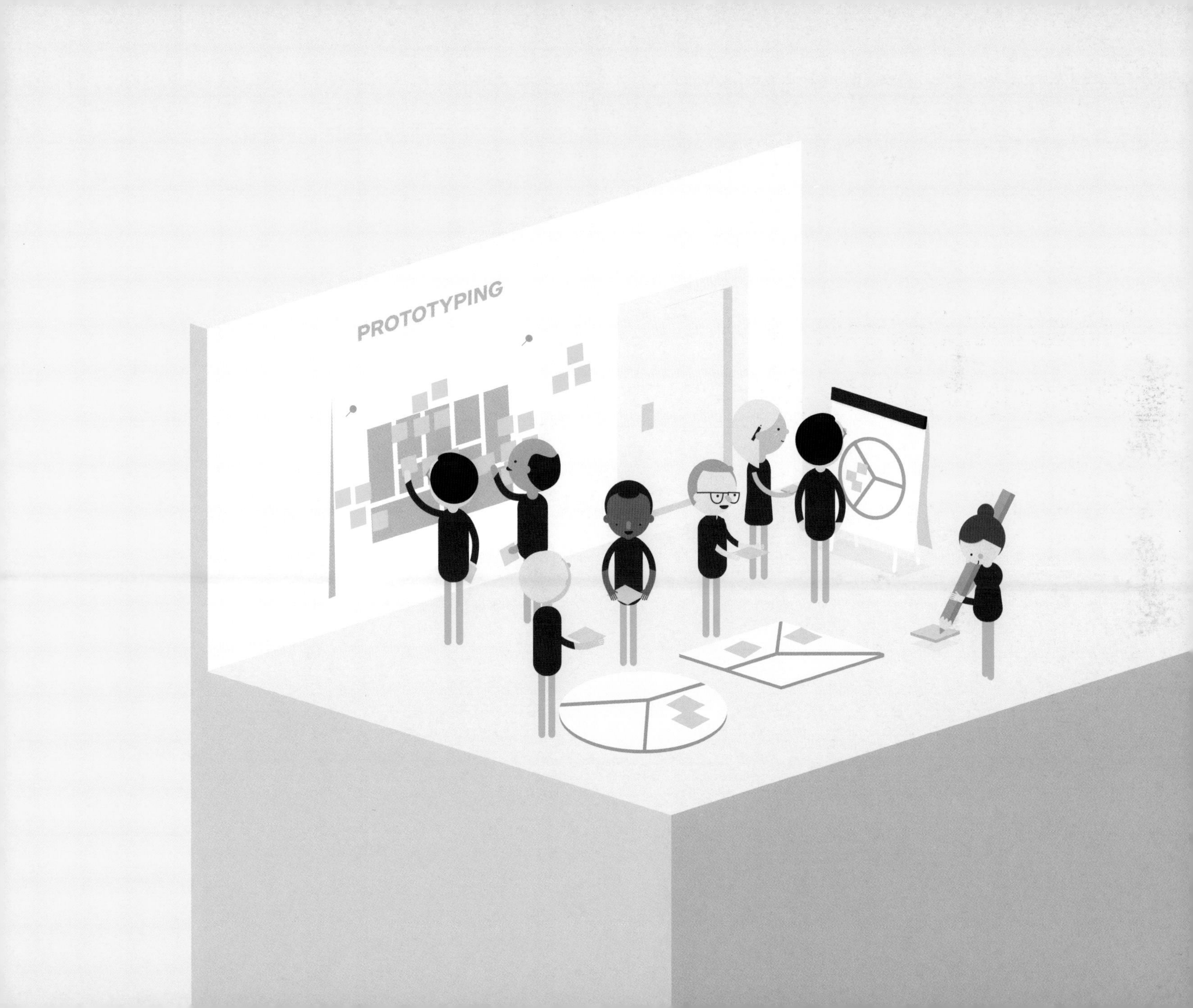
PROTOTYPING

What's Prototyping?

Use the activity of making quick and rough study models of your idea to explore alternatives, shape your value proposition, and find the best opportunities. Prototyping is common in the design professions for physical artifacts. We apply it to the concept of value propositions to rapidly explore possibilities before testing and building real products and services.

DEF·I·NI·TION

Prototyping

The practice of building quick, inexpensive, and rough study models to learn about the desirability, feasibility, and viability of alternative value propositions and business models.

Quickly explore radically different directions for the same idea with the following prototyping techniques before refining one in particular.

Napkin Sketches

p. 80

Make alternatives tangible with napkin sketches. Use a single sketch for every potential direction your idea could take.

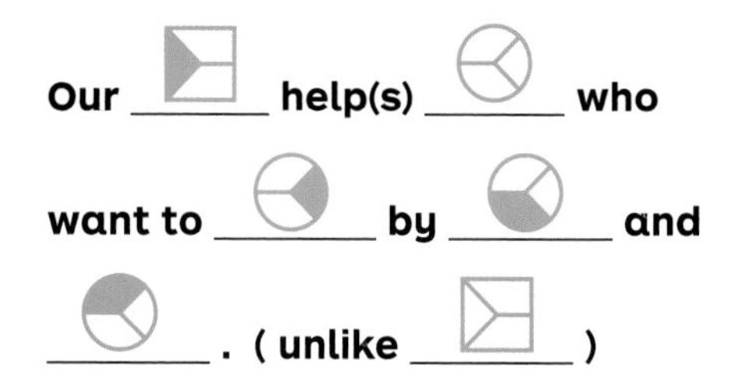

Ad-libs

p. 82

Pinpoint how different alternatives create value by filling in the blanks in short ad-libs.

Tips

- Spend a maximum of 5 to 15 minutes on sketching out your early prototypes.
- Always use a visible timer and stick to a predefined time frame.
- Don't discuss too long which one of several possible directions to prototype. Prototype several of them quickly and then compare.
- Remember constantly that prototyping is an exploratory tool. Don't spend time on the details of a prototype that is likely to change radically anyway.

Value Proposition Canvases

➔ p. 84

Flesh out possible directions with the Value Proposition Canvas. Understand which jobs, pains, and gains each alternative is addressing.

Representation of a Value Proposition

➔ p. 234

Help customers and partners understand potential value propositions by bringing them to life without building them.

Minimum Viable Product

➔ p. 222

Build a minimum feature set that brings your value proposition to life and allows testing it with customers and partners.

More in section 3. Test, ➔ p. 172

10 Prototyping Principles

Unlock the power of prototyping. Resist the temptation of spending time and energy refining one direction only. Rather, use the principles described here to explore multiple directions with the same amount of time and energy. You will learn more and discover better value propositions.

Get "10 Prototyping Principles" poster

1

Make it visual and tangible.

These kinds of prototypes spark conversations and learning. Don't regress into the land of blah blah blah.

2

Embrace a beginner's mind.

Prototype "what can't be done." Explore with a fresh mind-set. Don't let existing knowledge get in the way of exploration.

3

Don't fall in love with first ideas—create alternatives.

Refining your idea(s) too early prevents you from creating and exploring alternatives. Don't fall in love too quickly.

4

Feel comfortable in a "liquid state."

Early in the process the right direction is unclear. It's a liquid state. Don't panic and solidify things too early.

5

Start with low fidelity, iterate, and refine.

Refined prototypes are hard to throw away. Keep them rough, quick, and cheap. Refine with increasing knowledge about what works and what doesn't.

6

Expose your work early —seek criticism.

Seek feedback early and often before refining. Don't take negative feedback personally. It's worth gold to improve your prototype.

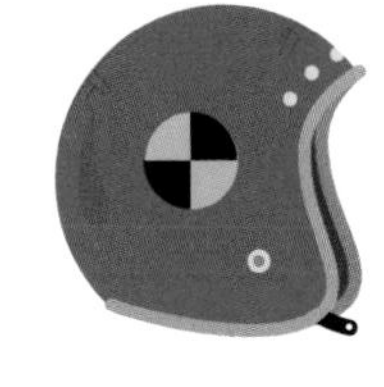

7

Learn faster by failing early, often, and cheaply.

Fear of failure holds people back from exploring. Overcome that with a culture of rough and quick prototyping that keeps failure cheap and leads to faster learning.

8

Use creativity techniques.

Use creativity techniques to explore groundbreaking prototypes. Dare to break out of how things are usually done in your company or industry.

9

Create "Shrek models."

Shrek models are extreme or outrageous prototypes that you are unlikely to build. Use them to spark debate and learning.

10

Track learnings, insights, and progress.

Keep track of all your alternative prototypes, learnings, and insights. You might use earlier ideas and insights later in the process.

Make Ideas Visible with Napkin Sketches

OBJECTIVE
Quickly visualize ideas for value propositions

OUTCOME
Alternative prototypes in the form of napkin sketches

Napkin sketches are a rough representation of a value proposition or business model and highlight only the core idea, not how it works. They are rough enough to fit on the back of a napkin and still communicate the idea. Use them early in your prototyping process to explore and discuss alternatives.

What is a napkin sketch?
Napkin sketches are a cheap way to make your ideas more tangible and shareable. They avoid going into the details of how an idea works to steer clear of getting hung up with implementation issues.

What is it used for?
Use napkin sketches to quickly share and evaluate ideas during the early value proposition design process. Their roughness is deliberate so you can throw ideas away without regret and explore alternatives. You may also use them to gather early feedback from customers.

Caveat
Make sure people understand that napkin sketches are an exploratory tool. You will kill or transform many of the sketched out ideas during the prototyping and testing process.

The best napkin sketches...

Contain only one core idea or direction (ideas can be merged later).

Explain what an idea is about, not how it will work (no processes or business models yet!).

Keep things simple enough to get it in a glance (details are for more refined prototypes later on).

Can be pitched in 10 to 30 seconds.

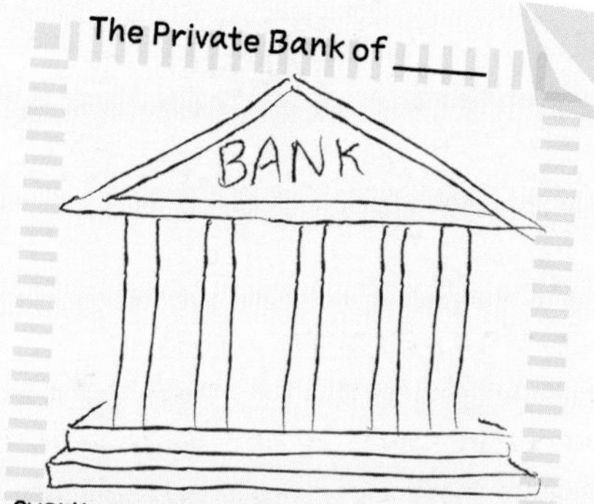

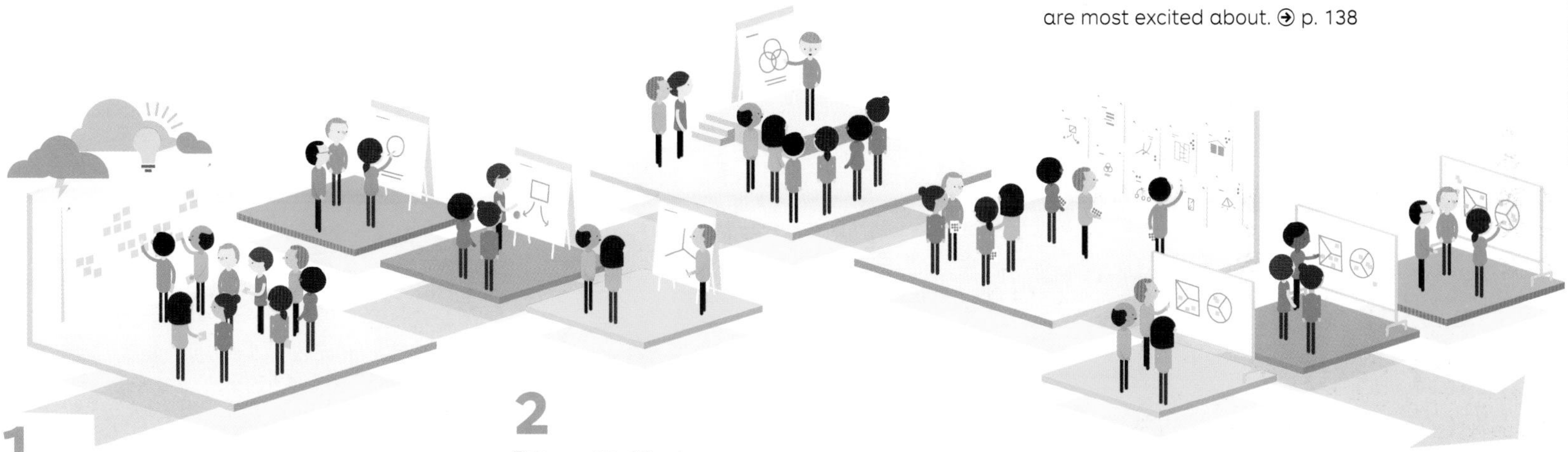

1

Brainstorm · **15–20 min**

Use different brainstorming techniques, such as trigger questions ➔ p. 15, 17, 31, 33 or "what if" questions to generate a large quantity of possible directions for interesting value propositions. Don't worry about choice at this stage. Quantity is better than quality. These are quick and dirty prototypes that will change inevitably.

2

Draw · **12–15 min**

Participants split into break-out groups, and each group quickly picks three ideas for three alternative value propositions. They draw a napkin sketch for each one on a flip chart. Making two or three sketches increases diversity and reduces the risk of endless discussions.

3

Pitch · **30 sec per group**

One team member of each break-out group takes the stage and pitches the (large) napkin sketches. Each pitch should be no longer than 30 seconds—just enough to outline what the idea is about, not to explain how it works! Make sure there is sufficient diversity across the groups or else send everybody back to the drawing board.

4

Display

All napkin sketches are exposed in a sort of gallery on the wall. You should now have a nice diversity of alternative directions.

5

Dotmocracy · **10–15 min (ideally over a break)**

Participants get 10 stickers to vote for their favorite ideas. They can give all votes to one idea or distribute them among several napkin sketches. This is not a decision-making mechanism. It is a process that highlights the ideas that participants are most excited about. ➔ p. 138

6

Prototype

Break-out groups continue by sketching out a value proposition canvas for the one napkin sketch out of their three that got most votes. Potentially redistribute the napkin sketches that got most votes among the different groups.

Create Possibilities Quickly with Ad-Libs

OBJECTIVE
Quickly shape potential value proposition directions

OUTCOME
Alternative prototypes in the form of "pitchable" sentences

Ad-libs are a great way to quickly shape alternative directions for your value proposition. They force you to pinpoint how exactly you are going to create value. Prototype three to five different directions by filling out the blanks in the ad-lib below.

Download the template

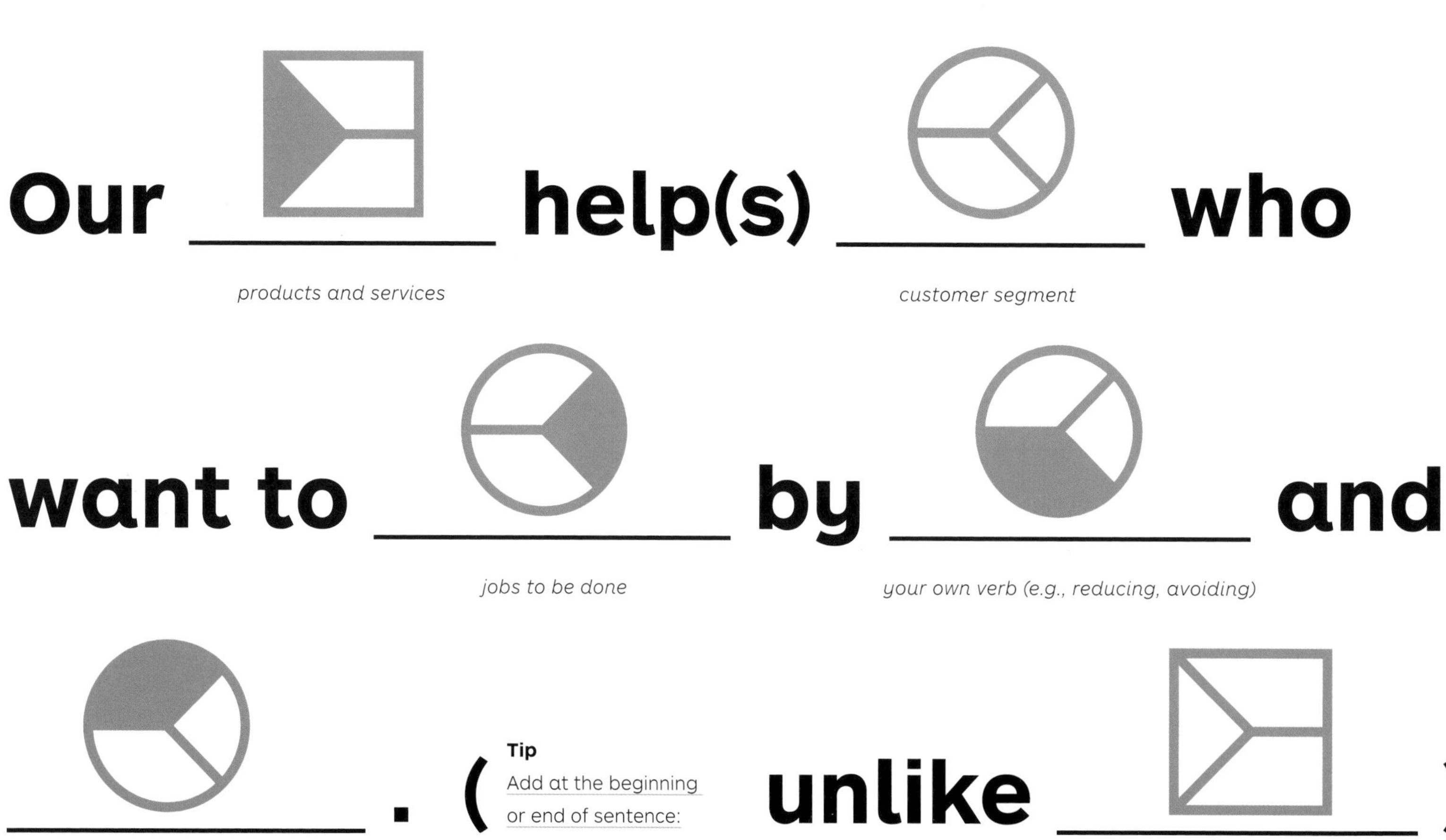

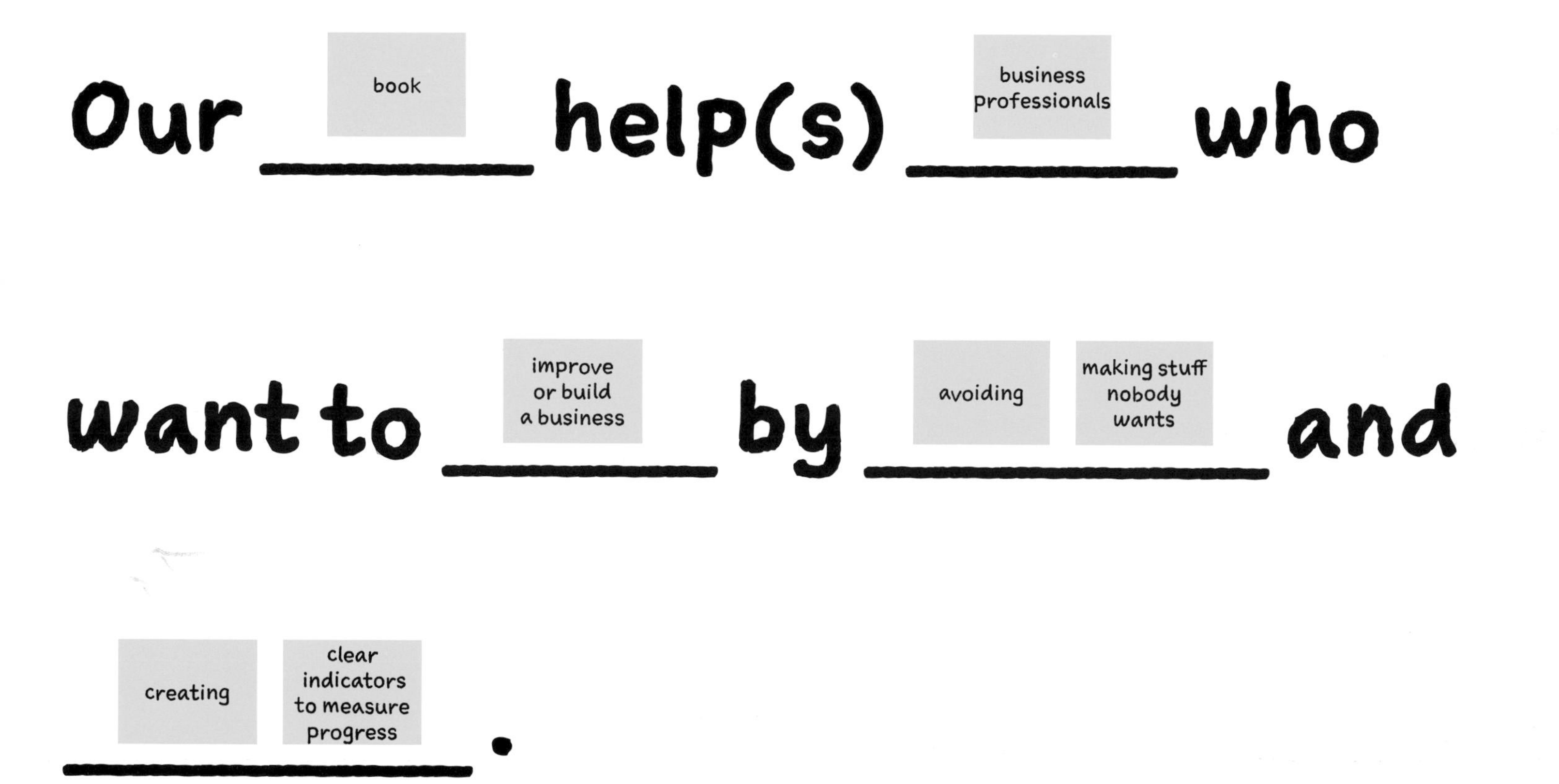
Our
book
help(s)
business professionals
who
want to
improve or build a business
by
avoiding
making stuff nobody wants
and
creating
clear indicators to measure progress
.

Flesh out Ideas with Value Proposition Canvases

OBJECTIVE
Sketch explicitly how different ideas create customer value

OUTCOME
Alternative prototypes in the form of Value Proposition Canvases

Use the Value Proposition Canvas to sketch out quick alternative prototypes, just like you would with napkin sketches or ad-libs. Don't just work with the canvas to refine final ideas, but use it as an exploratory tool until you find the right direction.

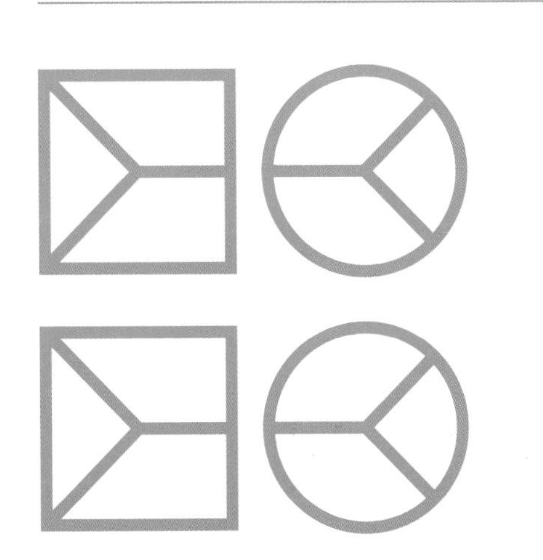

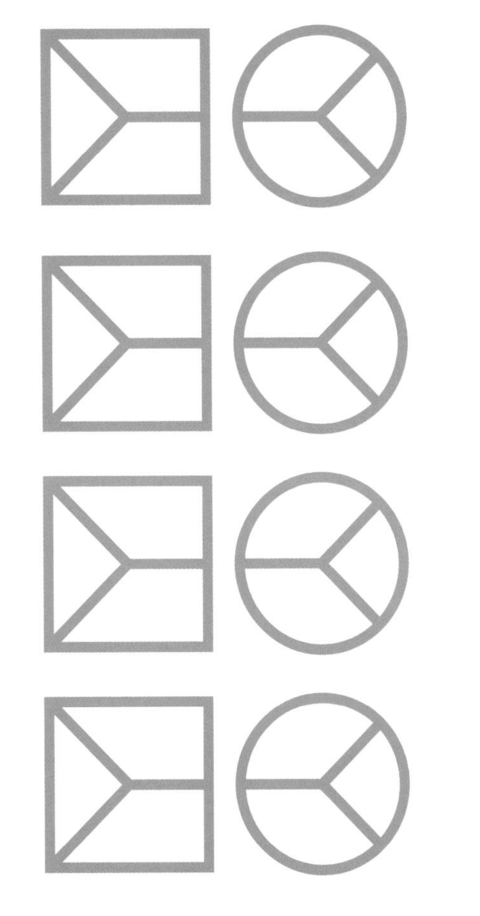

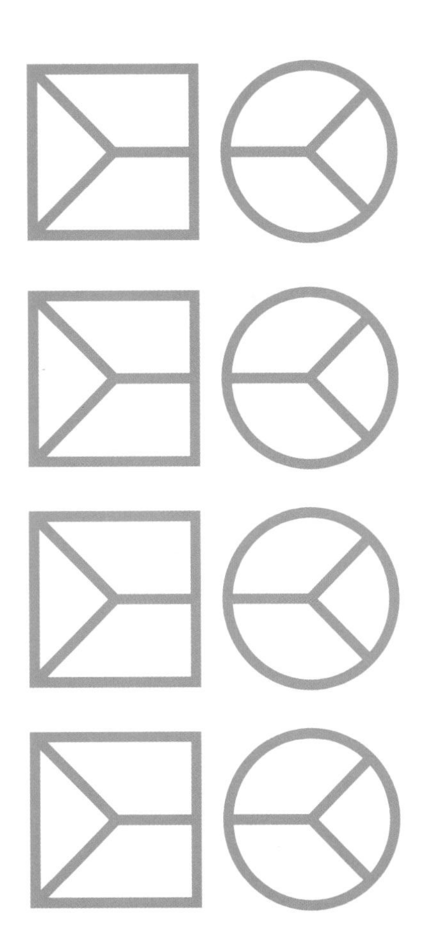

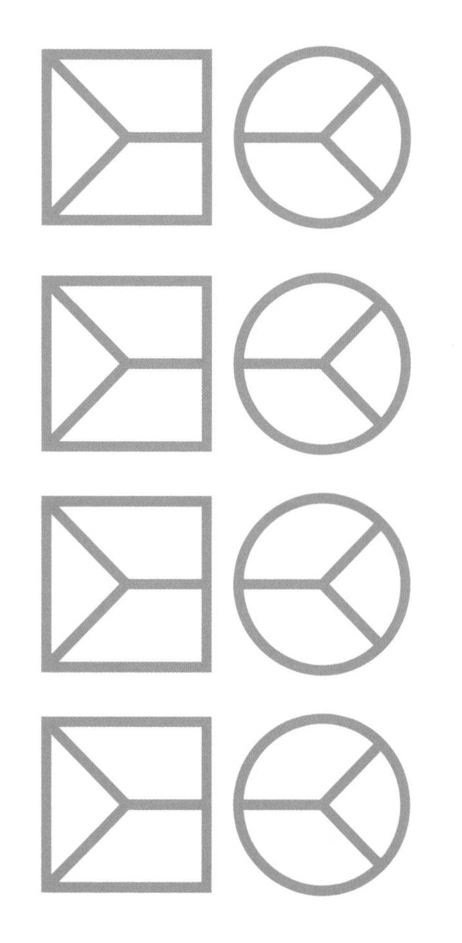

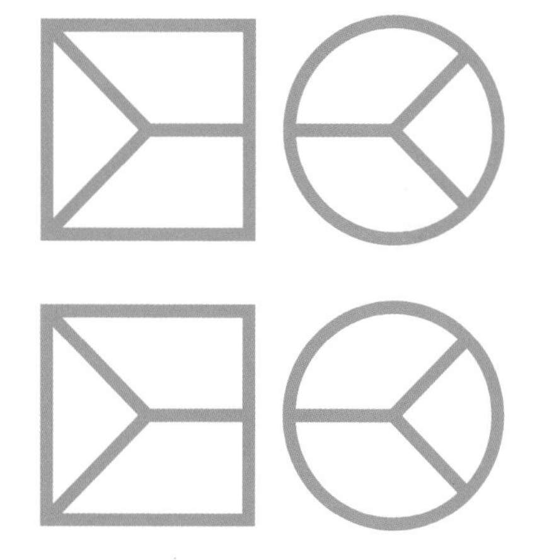

Use a visible timer to constrain the time you spend working on a specific prototype. Keep early prototypes short.

Don't be afraid to prototype radical directions, even if you know you are unlikely to pursue them. Explore and learn.

2.2 Starting Points

STARTING POINTS
Learning Card
Strategyzer
We believe that
We observed
From that we learned that
Therefore we will

Where to Start

Contrary to popular belief, great new value propositions don't always have to start with the customer. They do, however, always have to end with addressing jobs, pains, or gains that customers care about.

On this spread we offer 16 trigger areas to get started with new or improved value propositions. They start from either the customer, your existing value propositions, your business models, your environment, or business models and value propositions from other industries and sectors.

Get "Innovation Starting Points" poster

Could you...

Zoom out

Zoom in

Imitate and "import" a pioneering model from another sector or industry?

Create value based on a new technology trend or turn a new regulation to your advantage?

Come up with a new value proposition that your competitors can't copy?

Come up with a new value proposition based on a new partnership?

Build on your existing activities and resources, including patents, infrastructure, skills, user base?

Dramatically alter your cost structure to lower your prices substantially?

Create a new gain creator for a given customer profile?

Imagine a new product or service?

Create a new pain reliever for a given customer profile?

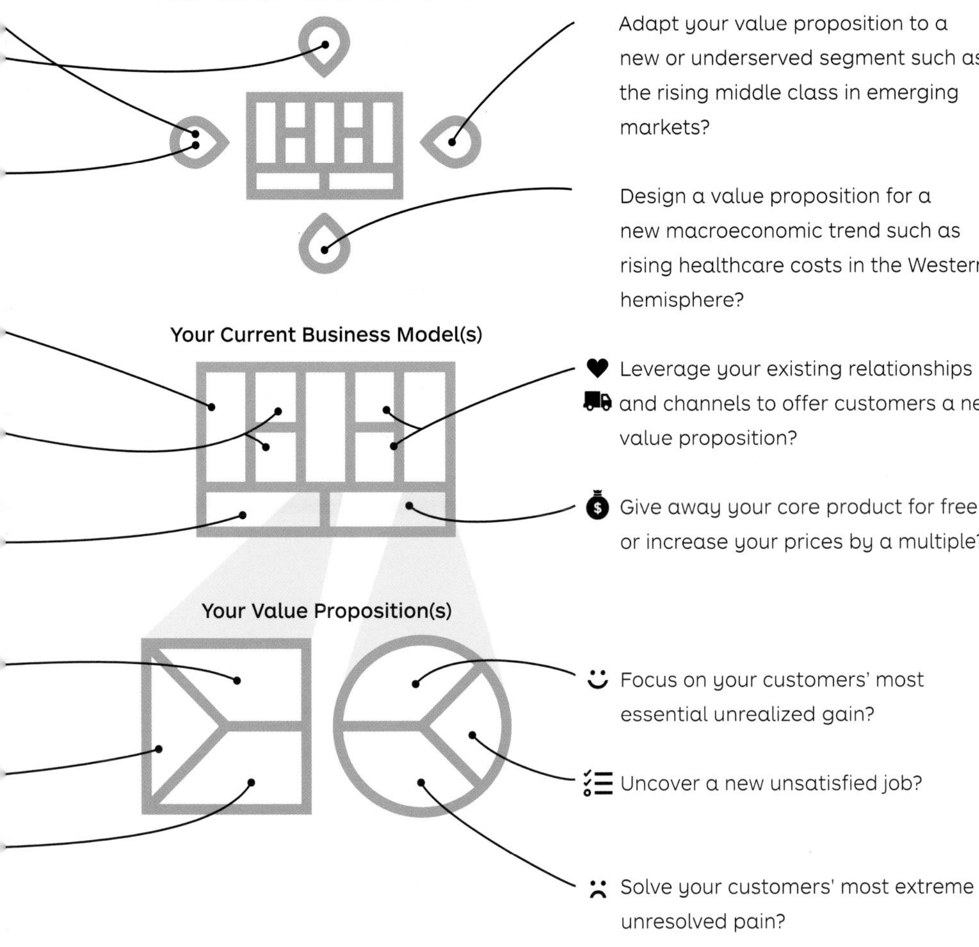

Adapt your value proposition to a new or underserved segment such as the rising middle class in emerging markets?

Design a value proposition for a new macroeconomic trend such as rising healthcare costs in the Western hemisphere?

Leverage your existing relationships and channels to offer customers a new value proposition?

Give away your core product for free or increase your prices by a multiple?

Focus on your customers' most essential unrealized gain?

Uncover a new unsatisfied job?

Solve your customers' most extreme unresolved pain?

Spark Ideas with Design Constraints

Use design constraints to force people to think about innovative value propositions embedded in great business models. We outline five constraints of businesses whose value proposition and business model you can copy into your own arena. Don't hesitate to come up with other ones.

OBJECTIVE
Force yourself to think outside of the box

OUTCOME
Ideas that differ from your "usual" value propositions and business models

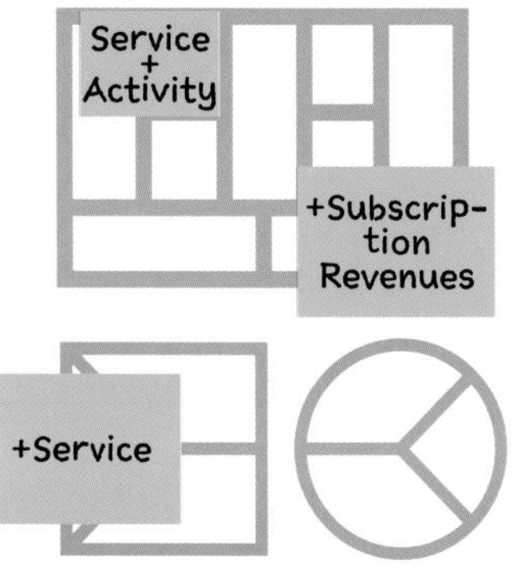

Servitization

Constraint: Transform from selling a product-based value proposition to a service-based one that generates revenues from a subscription model.

Hilti shifted from selling machine tools to builders to leasing fleet management services to managers at construction companies.

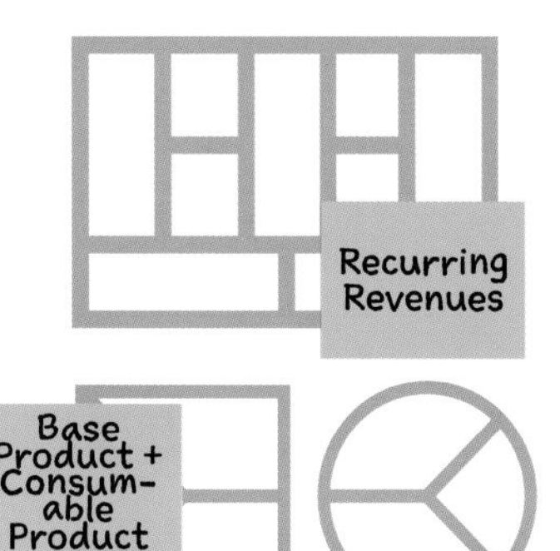

NESPRESSO

Razor Blade

Constraint: Create a value proposition composed of a base product and a consumable product that generates recurring revenues.

Nespresso transformed the sales of espresso from a transactional business to one with recurring revenues based on consumable pods for its espresso machine.

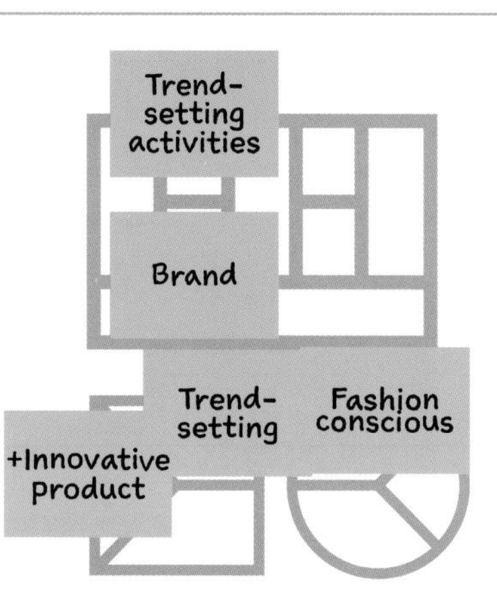

swatch

Trendsetter

Constraint: Transform a technology (innovation) into a fashionable trend.

Swatch conquered the world by turning a plastic watch that could be made cheaply due to a reduced number of pieces and innovative production technology into a global fashion trend.

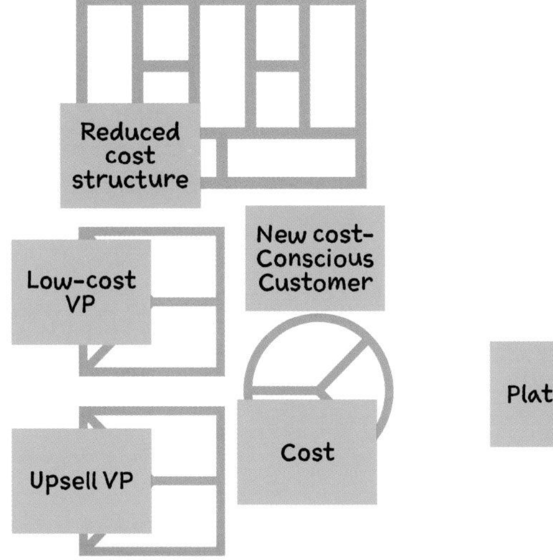

Low-Cost

Constraint: Reduce the core value proposition to its basic features, target an unserved or underserved customer segment with a low price and sell everything else as an additional value proposition.

Southwest became the largest low-cost airline by stripping down the value proposition to its bare minimum, travel from point A to point B, and offering low prices. They opened up flying to a new segment.

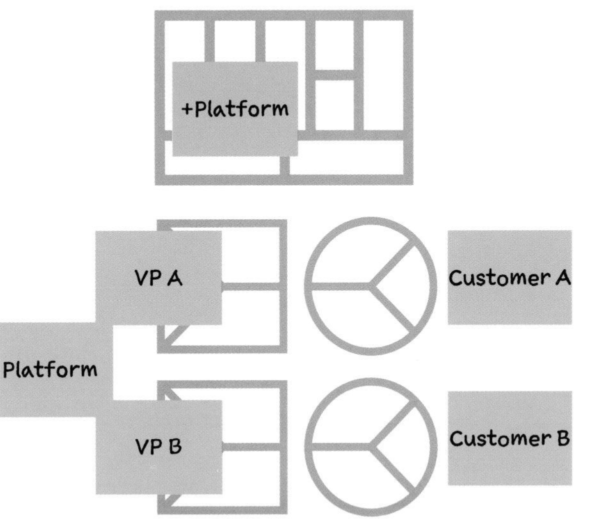

Platform

Constraint: Build a platform model that connects several actors with a specific value proposition for each.

Airbnb made private homes around the world accessible to travelers by connecting them with people who seek to rent out their apartments short term.

Tips

- Assign different constraints to different working groups if you have the opportunity to do so. It allows you to explore alternatives in parallel.
- Use constraints that represent the challenges in your arena, such as free value propositions, decreasing margins, and so on.

Download Constraint Cards

Invite Big Ideas to the Table with Books and Magazines

OBJECTIVE	OUTCOME
Broaden horizon and generate fresh ideas	Ideas that build on relevant topics and integrate latest trends

Use best-selling books and magazines to generate fresh ideas for new and innovative value propositions and business models. It's a quick and effective way to immerse yourself in various relevant and popular topics and build on current trends.

Bringing books into a workshop is like inviting the world's best thinkers to brainstorm. This way you can afford a lot more of them at the same time.

1

Select books.
Prepare a series of books and magazines representing a trend, important topic, or big idea on a large table. Ask workshop participants to pick up a book each.

E-retailers grow in power

Climate change awareness affects consumer behavior

Rise of the "sharing economy"

Mass-collab. changes how value is created

Meet and exceed the basic needs of every human being on the planet

How the digital generation ticks differently

Surge of the maker movement

2

Browse and extract.
Participants browse their book and capture the best ideas on sticky notes. (45 min)

Tips

- Select books about society, technology, and environment that push participants outside of their comfort zone.
- Avoid complicated business theories or methods.
- Mix in YouTube videos of keynote talks by the authors.
- Use napkin sketches to share your value proposition ideas.

3

Share and discuss.

Participants share their highlights in groups of four or five people and capture their insights on a board. (20 min)

4

Brainstorm possibilities.

Each group generates three new value proposition ideas based on their discussions. (30 min)

5

Pitch.

Each group shares their alternative value propositions with the other groups.

Download "Big Idea Book List"

Push vs. Pull

The push versus pull debate is a common one. *Push* indicates that you're starting the design of your value proposition from a technology or innovation you possess, whereas *pull* means you're beginning with a manifest customer job, pain, or gain. These are two common starting points, many of which we outlined previously ➔ p. 88. Consider both as viable options depending on your preferences and context.

Technology Push

Start from an invention, innovation, or (technological) resource for which you develop a value proposition that addresses a customer job, pain, and gain. In simple terms, this is a solution in search of a problem.

Explore value proposition prototypes that are based on your invention, innovation, or technological resource with potentially interested customer segments. Design a dedicated value map for each segment until you find problem-solution fit. Read more about the build, measure, learn cycle on ➔ p. 186.

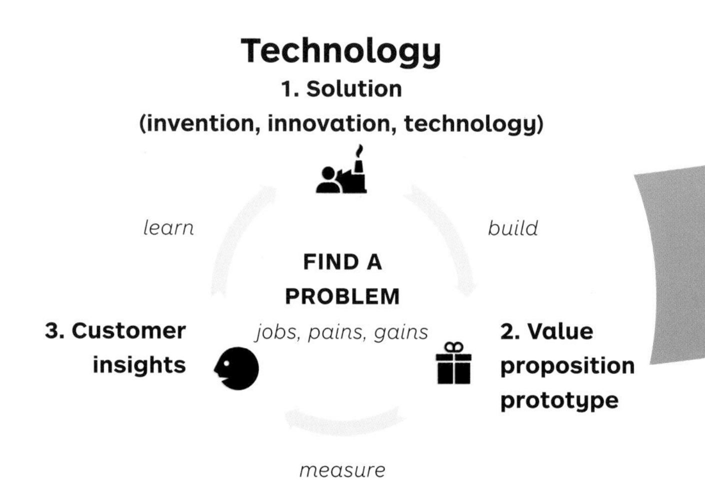

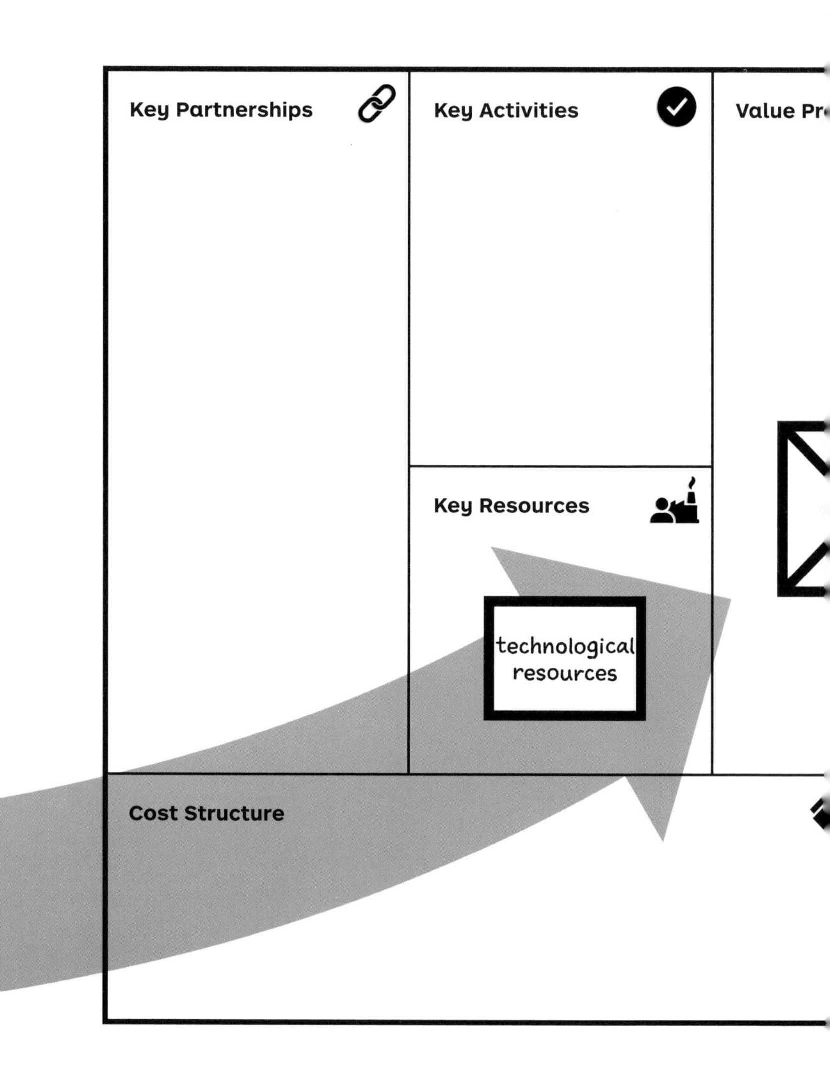

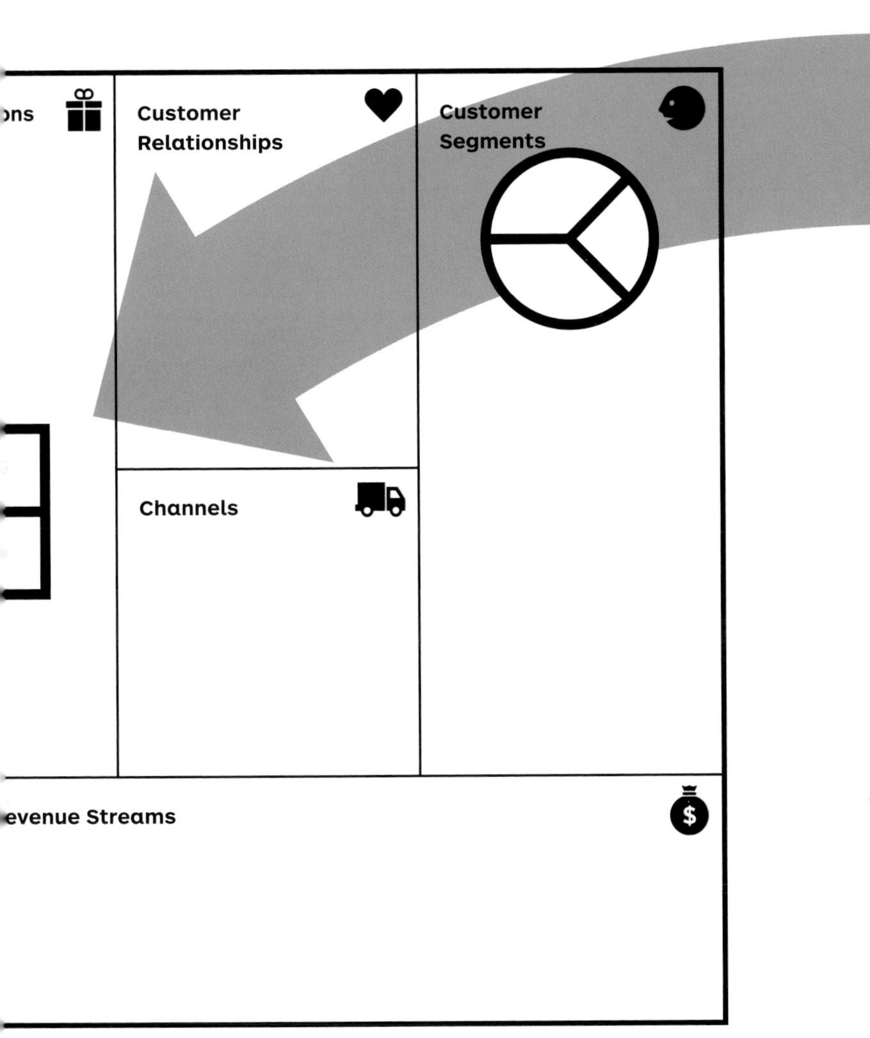

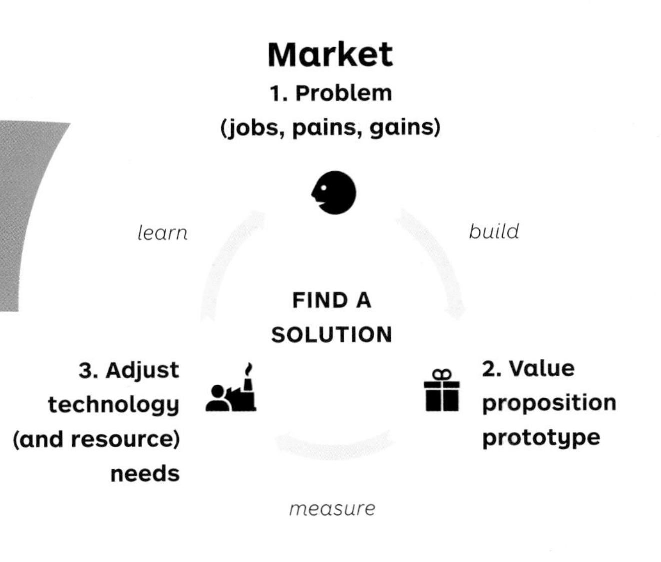

Market Pull

Start from a manifest customer job, pain, or gain for which you design a value proposition. In simple terms, this is a problem in search of a solution.

Learn what technologies and other resources are required for each value proposition prototype designed to address manifest customer jobs, pains, and gains. Redesign your value map and adjust resources until you find a viable solution to address customer jobs, pains, and gains. More about the build, measure, learn cycle on ➔ p. 186.

Push: Technology in Search of Jobs, Pains, and Gains

OBJECTIVE
Practice the technology-driven approach with no risk

OUTCOME
Improved skills

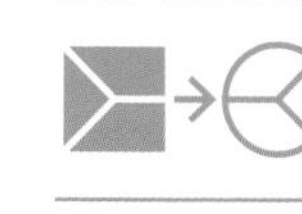

This push exercise starts with the solution

1
Design.
Design a value proposition based on the technology outlined in the press excerpt from the Swiss Federal Institute of Technology in Lausanne (EPFL) by targeting a customer segment that might be interested in adopting this technology.

2
Ideate.
Come up with an idea for a value proposition using the compressed air energy storage.

3
Segment.
Select a customer segment that could be interested in this value proposition and would be ready to pay for it.

The Business Model Canvas

Key Partnerships
Key Activities
Key Resources
compressed air energy storage
Value Propositions
?
Customer Relationships
Channels
Customer Segments
?
Cost Structure
Revenue Streams

Strategyzer

"Solar and aeolian sources are great candidates for the electricity generation of the future... However, solar and wind sources' peak availability takes place at times that do not usually correspond to peak demand hours. Therefore, a way must be devised to store and later reuse the energy generated.

EPFL has worked for over ten years on an original storage system: compressed air. The use of a hydraulic piston delivers the best system performance... The obtained high pressure air can be safely stored in bottles without losses until it is necessary to generate new electricity by expanding the gas in the cylinder. One of the advantages of our system is that it does not require rare materials.

A spin-off has been created to develop this principle and create 'turnkey' electrical energy storage and retrieval units. In 2014, a 25 kW pilot will be installed at a photovoltaic park in Jura.... In the future, there will be 250 kW installations at first and 2,500 kW ones afterwards."

Tips

- Add design constraints to technology push exercises. Your organization might not want to address certain customer segments (e.g., B2B, business-to-consumer [B2C], specific regions, etc.). Or you might prefer certain strategic directions, for example, licensing rather than building solutions.
- Follow up on your customer assumptions by researching customers ➔ p. 104 and producing evidence ➔ p. 172 once you've selected a potentially interested segment.

Zoom in

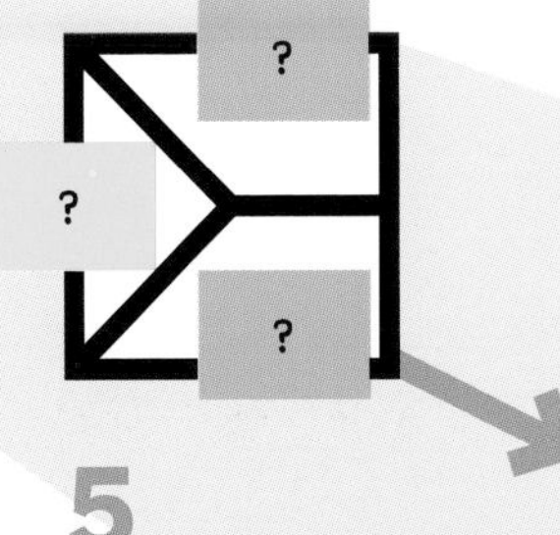

4

Profile.
Sketch out the customer's profile. Make assumptions about jobs to be done, pains, and gains.

5

Sketch.
Refine the value proposition by sketching out how it will kill customer pains and create gains.

6

Assess.
Assess the fit between the customer profile and the designed value proposition.

Continued on ➔ p. 152

The Value Proposition Canvas

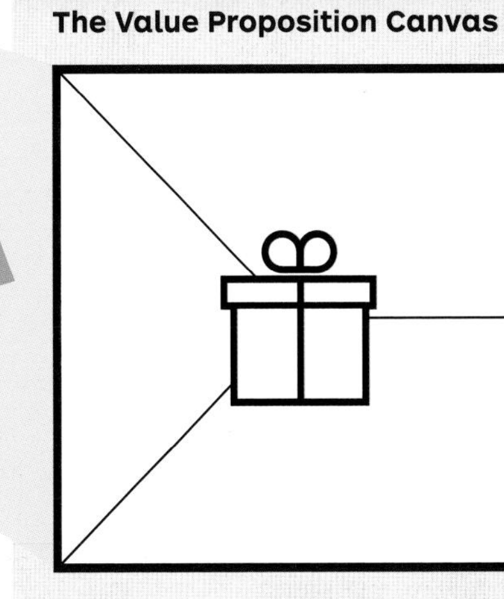

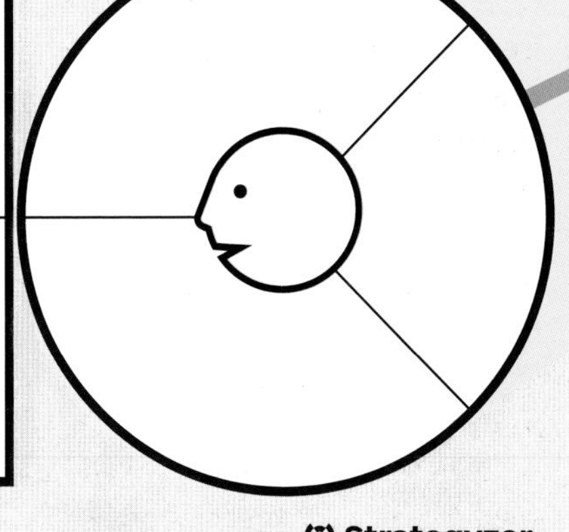

Pull: Identify High-Value Jobs

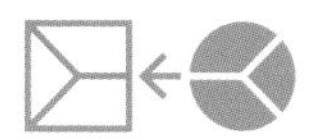

Great value proposition creators master the art of focusing on the jobs, pains, and gains that matter. How will you know which of these jobs, pains, and gains to focus on? Identify high-value jobs by asking if they are important, tangible, unsatisfied, and lucrative.

!

+

+

High-value jobs are characterized by pains and gains that are...

Important

When the customer's success or failure to get the job done leads to essential gains or extreme pains, respectively.

- Does failing the job lead to extreme pains?
- Does failing the job lead to missing out on essential gains?

Tangible

When the pains or gains related to a job can be felt or experienced immediately or often, not just days or weeks later.

- Can you feel the pain?
- Can you see the gain?

Unsatisfied

When current value propositions don't help relieve pains or create desired gains in a satisfying way or simply don't exist.

- Are there unresolved pains?
- Are there unrealized gains?

Lucrative

When many people have the job with related pains and gains or a small number of customers are willing to pay a premium.

- Are there many with this job, pain, or gain?
- Are there few willing to pay a lot?

High-value jobs

Focus on the highest-value jobs and related pains and gains.

Based on initial work by consultancy, Innosight.

Pull: Job Selection

OBJECTIVE
Identify high-value customer jobs that you could focus on

OUTCOME
Ranking of customer jobs from your perspective

This pull exercise starts with the customer.

Imagine your customers are chief information officers (CIOs) and you have to understand which jobs matter most to them. Do this exercise to prioritize their jobs or apply it to one of your own customer profiles.

Tips

- This exercise helps you prioritize jobs from the customer's perspective. It doesn't mean you have to mandatorily address the most important ones in your value proposition; those might be outside your scope. However, make sure your value proposition does address jobs that are highly relevant to customers.
- Great value proposition creators often focus on only few jobs, pains, and gains, but do that extremely well.
- Complement this exercise with getting customer insights from the field ➔ p. 106 and experiments that produce evidence ➔ p. 216.

Customer Profile
Synthesized customer profile of a CIO

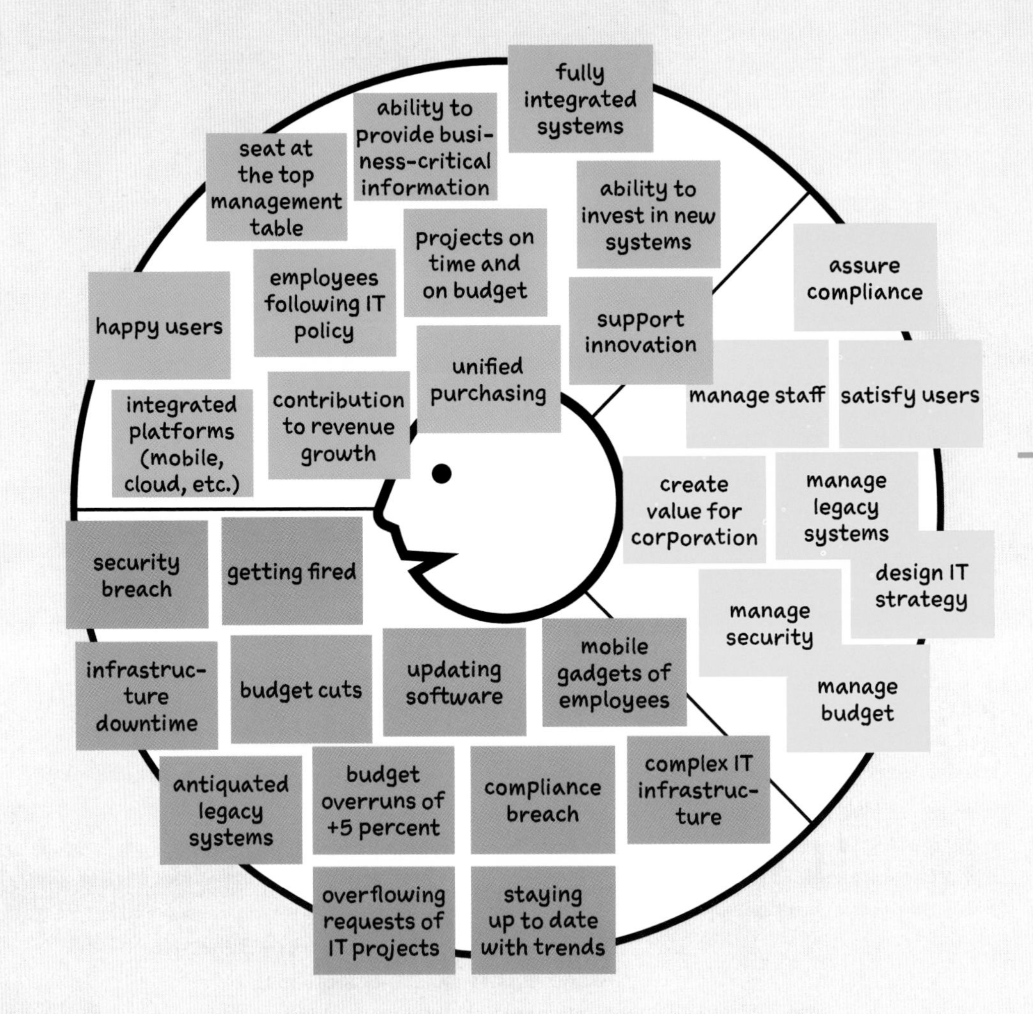

Strategyzer

Copyright Business Model Foundry AG
The makers of Business Model Generation and Strategyzer

- Does failing the job lead to extreme pains?
- Does failing the job lead to missing out on essential gains?

- Can you feel the pain?
- Can you see the gain?

- Are there unresolved pains?
- Are there unrealized gains?

- Are there many with this job, pain, or gain?
- Are there few willing to pay a lot?

Focus on the highest value jobs and related pains and gains.

Jobs	Important	Tangible	Unsatisfied	Lucrative	High-value jobs
create value for corporation	•••	•	•••	••	= 9
design IT strategy	••	•	••	••	= 7

Scoring scale: • *(low) to* •••• *(high)*

Based on initial work by consultancy, Innosight.

Six Ways to Innovate from the Customer Profile

You've mapped your customer profile. What to do from here? Here are six ways to trigger your next value proposition move.

Can you...

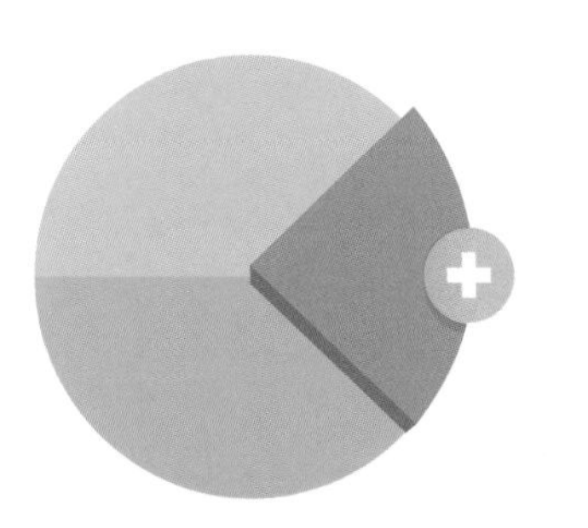

Address more jobs?

Address a more complete set of jobs, including related and ancillary jobs.

With the iPhone, Apple not only reinvented the mobile phone but enabled us to store and play music and browse the web on one device.

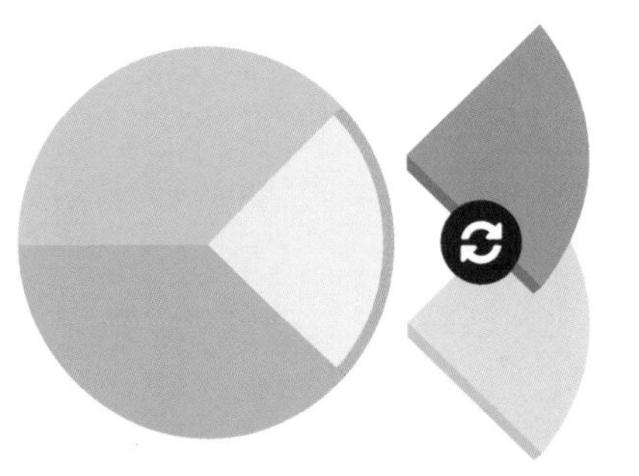

Switch to a more important job?

Help customers do a job that is different from what most value propositions currently focus on.

Hilti, the machine tool manufacturer, understood that construction managers needed to keep schedules to avoid penalties, not only drill holes. Their fleet management solution addressed the former in addition to the latter.

Go beyond functional jobs?

Look beyond functional jobs and create new value by fulfilling important social and emotional jobs.

Mini Cooper created a car that became as much a means of transport as a statement of identity.

Download trigger questions

Help a lot more customers get a job done?

Help more people do a job that was otherwise too complex or too expensive.

High-end web data storage and computing power used to be reserved to big companies with large IT budgets. Amazon.com made it available to companies of any size and budget with Amazon Web Services.

Get a job done incrementally better?

Help customers better do a job by making a series of microimprovements to an existing value proposition.

German engineering and electronics multinational Bosch improved on a wide range of features of its circular saw that really mattered to customers and outperformed competition.

Help a customer get a job done radically better?

This is the stuff of new market creation, when a new value proposition dramatically outperforms older ways of helping a customer get a job done.

The first spreadsheet called VisiCalc not only introduced a new market for such tools but also ushered a whole new realm of possibilities across industries powered by easy, visual calculations.

2.3

Understanding Customers

Acme
OBSERVATION

Six Techniques to Gain Customer Insights

Understanding the customer's perspective is crucial to designing great value propositions. Here are six techniques that will get you started. Make sure you use a good mix of these techniques to understand your customers deeply.

The Data Detective
Build on existing work with (desk) research. Secondary research reports and customer data you might already have provide a great foundation for getting started. Look also at data outside your industry and study analogs, opposites, or adjacencies.
Difficulty level: ★
Strength: great foundation for further research
Weakness: static data from a different context
➔ p. 108 for more

The Journalist
Talk to (potential) customers as an easy way to gain customer insights. It's a well-established practice. However, customers might tell you one thing in an interview but behave differently in the real world.
Difficulty level: ★★
Strength: quick and cheap to get started with first learnings and insights
Weakness: customers don't always know what they want and actual behavior differs from interview answers
➔ p. 110 for more

The Anthropologist
Observe (potential) customers in the real world to get good insights into how they really behave. Study which jobs they focus on and how they get them done. Note which pains upset them and which gains they aim to achieve.
Difficulty level: ★★★
Strength: data provide unbiased view and allow discovering real-world behavior
Weakness: difficult to gain customer insights related to new ideas
➔ p. 114 for more

The Impersonator

"Be your customer" and actively use products and services. Spend a day or more in your customer's shoes. Draw from your experience as an (unsatisfied) customer.

Difficulty level: ★★

Strength: firsthand experience of jobs, pains, and gains

Weakness: not always representative of your real customer or possible to apply

The Cocreator

Integrate customers into the process of value creation to learn with them. Work with customers to explore and develop new ideas.

Difficulty level: ★★★★★

Strength: the proximity with customers can help you gain deep insights

Weakness: may not be generalized to all customers and segments

The Scientist

Get customers to participate (knowingly or unknowingly) in an experiment. Learn from the outcome.

Difficulty level: ★★★★

Strength: provides fact-based insights on real-world behavior; works particularly well for new ideas

Weakness: can be hard to apply in existing organizations because of strict (customer) policies and guidelines

➔ p. 216 for more

The Data Detective: Get Started with Existing Information

Never before have creators had more access to readily available information and data inside and outside their companies before even getting started with designing value. Use available data sources as a launching pad to getting started with customer insights.

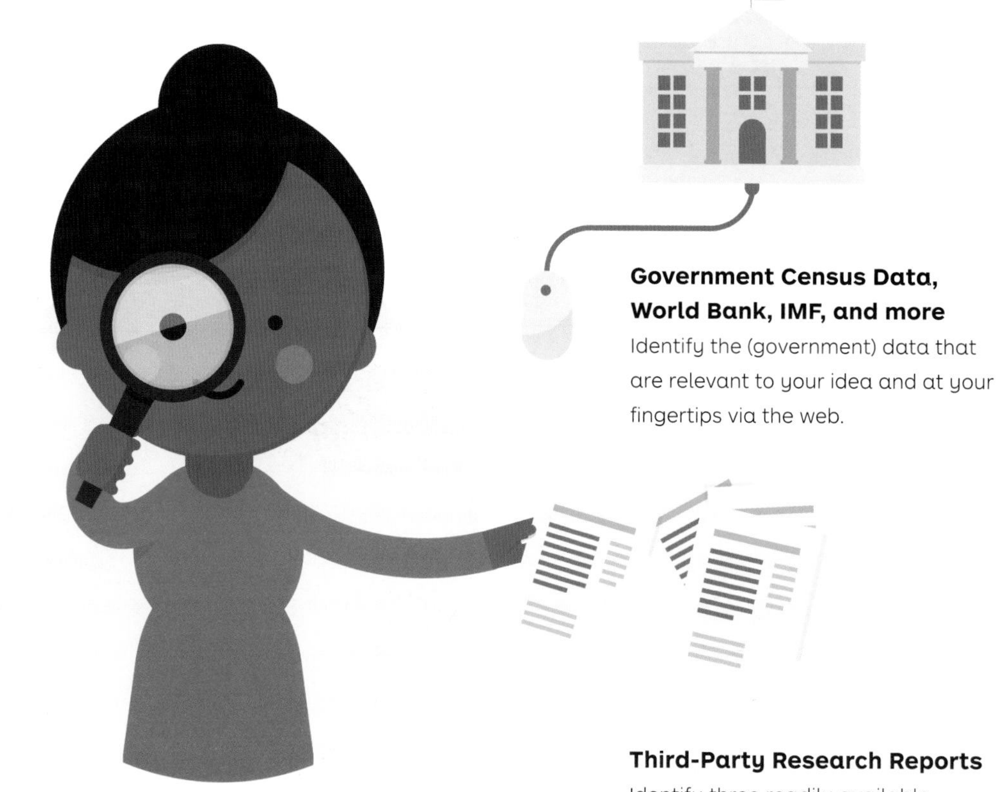

Google Trends
Compare three search terms representing three different trends related to your idea.

Government Census Data, World Bank, IMF, and more
Identify the (government) data that are relevant to your idea and at your fingertips via the web.

Google Keyword Planner
Learn what's popular with potential customers by finding the top five search terms related to your idea. How often are they searched for?

Third-Party Research Reports
Identify three readily available research reports that can serve you as a starting point to prepare your own customer and value proposition research.

Social Media Analytics

Existing companies and brands should:

- Identify the shakers and movers related to their brand on social media?
- Spot the 10 most frequently mentioned positive and negative things said about them on social media.

Customer Relationship Management (CRM)

- List the top three questions, complaints, and requests that you are getting from your daily interactions with customers (e.g., support).

Tracking Customers on Your Website

- List the top three ways customers reach your site (e.g., search, referrals).
- Find the 10 most and least popular destinations on your website.

Data Mining

Existing company should mine their data to:

- Identify three patterns that could be useful to their new idea.

Source: Siegel & Davenport, Predictive Analytics: The Power to Predict Who Will Click, Buy, Lie, or Die, 2013.

EXERCISE

The Journalist: Interview Your Customers

OBJECTIVE
Gain a better customer understanding

OUTCOME
First lightly validated customer profile(s)

Talk to customers to gain insights relevant to your context. Use the Value Proposition Canvas to prepare interviews and organize the chaotic mass of information that will be coming at you during the interview process.

1

Create a customer profile.
Sketch out the jobs, pains, and gains you believe characterize the customer you are targeting. Rank jobs, pains, and gains in order of importance.

2

Create an interview outline.
Ask yourself what you want to learn. Derive the interview questions from your customer profile. Ask about the most important jobs, pains, and gains.

5

Review the interview.
Assess if you need to review the interview questions based on what you learned.

3

Conduct the interview.

Conduct the interview by following the interview ground rules outlined on the next page.

4

Capture.

Map out the jobs, pains, and gains you learned about in the interview on an empty customer profile.

Make sure you also capture business model learnings. Write down your most important insights.

6

Search for patterns.

Can you discover similar jobs, pains, and gains? What stands out? What is similar or different among interviewees?

Why are they similar or different? Can you detect specific (recurring) contexts that influence jobs, pains, and gains?

7

Synthesize.

Make a separate synthesized customer profile for every customer segment that emerges from all your interviews. Write down your most important insights on sticky notes.

Tip

Capture your biggest insights from all the interviews.

Ground Rules for Interviewing

It is an art to conduct good interviews that provide relevant insights for value proposition design. Make sure you focus on unearthing what matters to (potential) customers rather than trying to pitch them solutions. Follow the rules on this spread to conduct great interviews.

Get "Ground Rules for Interviewing" poster

Rule 1

Adopt a beginner's mind.

Listen with a "fresh pair of ears" and avoid interpretation. Explore unexpected jobs, pains, and gains in particular.

Rule 2

Listen more than you talk.

Your goal is to listen and learn, not to inform, impress, or convince your customer of anything. Avoid wasting time talking about your own beliefs, because it's at the expense of learning about your customer.

Rule 3

Get facts, not opinions.

Don't ask, "Would you...?"
Ask, "When is the last time you have...?"

Rule 4

Ask "why" to get real motivations.

Ask, "Why do you need to do...?"
Ask, "Why is ____ important to you?"
Ask, "Why is ____ such a pain?"

Rule 5

The goal of customer insight interviews is not selling (even if a sale is involved); it's about learning.

Don't ask, "Would you buy our solution?" Ask "what are your decision criteria when you make a purchase of...?"

Rule 6

Don't mention solutions (i.e., your prototype value proposition) too early.

Don't explain, "Our solution does..."
Ask, "What are the most important things you are struggling with?"

Rule 7

Follow up.

Get permission to keep your interviewee's contact information to come back for more questions and answers or testing prototypes.

Rule 8

Always open doors at the end.

Ask, "Who else should I talk to?"

Tips

- Interviews are an excellent starting point to learn from customers, but typically they don't provide enough or sufficiently reliable insights for making critical decisions. Complement your interviews with other research, just like a good journalist does further research to find the real story behind what people tell. Add real-world observations of customers and experiments that produce hard data to your research mix.
- Conduct interviews in teams of two people. Decide in advance who will lead the interview and who will take notes. Use a recording device (photo, video, or other) if possible, but be aware that interviewees might not answer the same way with a recording device on the table.

Fitzpatrick, The Mom Test, 2013.

The Anthropologist: Dive into Your Customer's World

Dive deep into your (potential) customers' worlds to gain insights about their jobs, pains, and gains. What customers do on a daily basis in their real settings often differs from what they believe they do or what they will tell you in an interview, survey, or focus group.

B2C: Stay with the family.
Stay at one of your potential customers' home for several days and live with the family. Participate in daily routines. Learn about what drives that person.

B2C: Observe shopping behavior.
Go to a store where your (potential) customers shop and observe people for 10 hours. Watch. Can you detect any patterns?

B2B: Work alongside/consult.
Spend time working with or alongside a (potential) customer (e.g., in a consulting engagement). Observe. What keeps the person up at night?

B2B/B2C?
How could you immerse yourself in your (potential) customer's life? Be creative! Go beyond the usual boundaries.

B2C: Shadow your customer for a day.
Be your (potential) customer's shadow and follow him or her for a day. Write down all the jobs, pains, and gains you observe. Time stamp them. Synthesize. Learn.

A Day in the Life Worksheet

OBJECTIVE
Understand your customer's world in more detail
OUTCOME
Map of your customer's day

Capture the most important jobs, pains, and gains of the customer you shadowed.

Tips

- Observe and take notes. Hold back with interpretation based on your own experience. Stay nonjudgmental! Work like an anthropologist and watch with "fresh" eyes and an open mind-set.
- Pay attention to both what you see and what you don't see.
- Capture not only what you can observe but also what is not talked about such as feelings or emotions.
- Develop customer empathy as a critical mind-set to perform this type of contextual inquiry effectively.

Time	Activity (what I see)		Notes (what I think)
7 pm	brush kid's teeth before bed		parents annoyed by water splashing everywhere

Download "A Day in the Life" worksheet

Identify Patterns in Customer Research

OBJECTIVE	OUTCOME
Crystallize your customer	Synthesized Customer Profile(s)

Analyze your data and try to detect patterns once you have a good amount of customer research gathered. Search for customers with similar jobs, pains, or gains or customers that care about the same jobs, pains, or gains and make separate customer profiles.

1

Display.
Display all the customer profiles from your research on a large wall.

2

Group and segment.
Group similar customer profiles in to one or more separate segments if you can identify patterns in the jobs, pains, and gains.

3

Synthesize.
Synthesize the profiles from each segment into a single master profile. Identify the most common jobs, pains, and gains and use separate labels to describe them in the master profile.

4

Design.
Get started with prototyping value propositions after finishing your first attempt at customer segmentation. Design one or more value proposition prototypes with confidence, based on the newly identified patterns in the master profile.

Synthesis Example: Master profile of a business professional / book reader

To establish a master profile of book readers, we looked at the jobs, pains, and gains of the different customer profiles from our interviews. We synthesized the most frequent ones into the master profile by using representative labels.

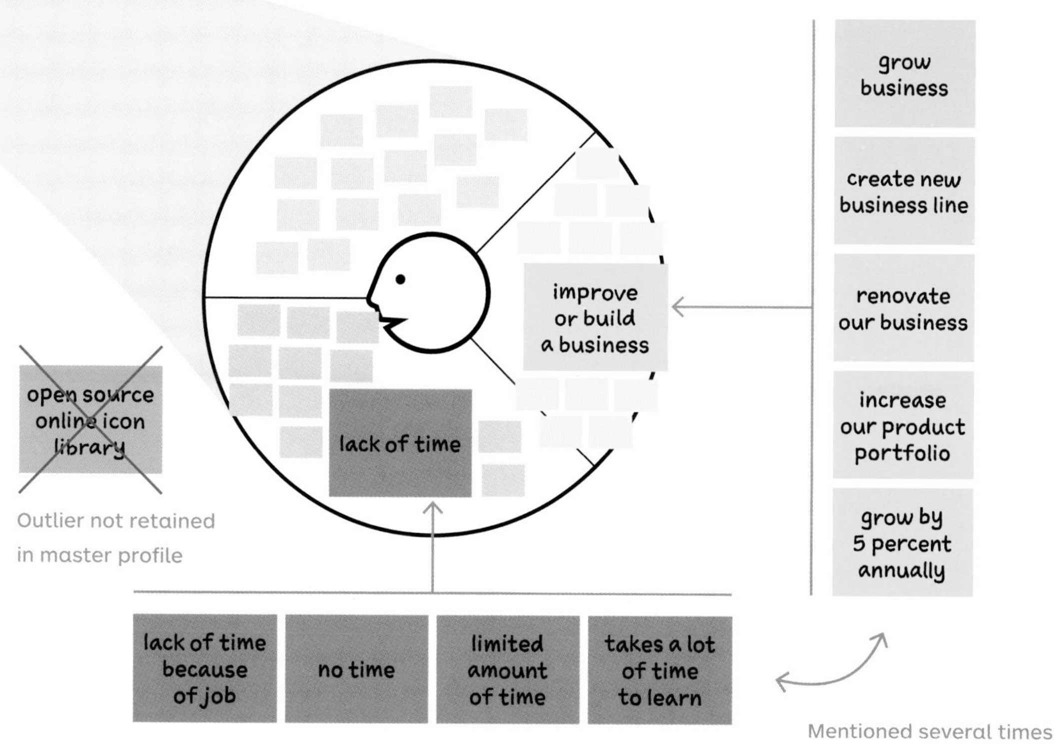

Outlier not retained in master profile

Mentioned several times

Tips

- Pay special attention to outlier profiles. They might be irrelevant, but they could represent a special learning opportunity. Sometimes the best discoveries lie at the edges.
- Ask yourself if an outlier might be a bellwether and a sign of things to come that you should pay attention to. Or maybe an outlier is different by positive deviance. It may simply be a better solution to jobs, pains, and gains than peers offer.

Find Your Earlyvangelist

Pay attention to earlyvangelists when researching potential customers and looking for patterns. The term was coined by Steve Blank* to describe customers who are willing and able to take a risk on a new product or service. Use earlyvangelists to build a foothold market and shape your value propositions via experimentation and learning.

5

Has or can acquire a budget.
The customer has committed or can quickly acquire a budget to purchase a solution.

4

Has put together solution out of piece parts.
The job is so important that the customer has cobbled together an interim solution.

3

Is actively looking for a solution.
The customer is searching for a solution *and* has a timetable for finding it.

2

Is aware of having a problem.
The customer understands that there is a problem or job.

1

Has a problem or need.
In other words, there is a job to be done.

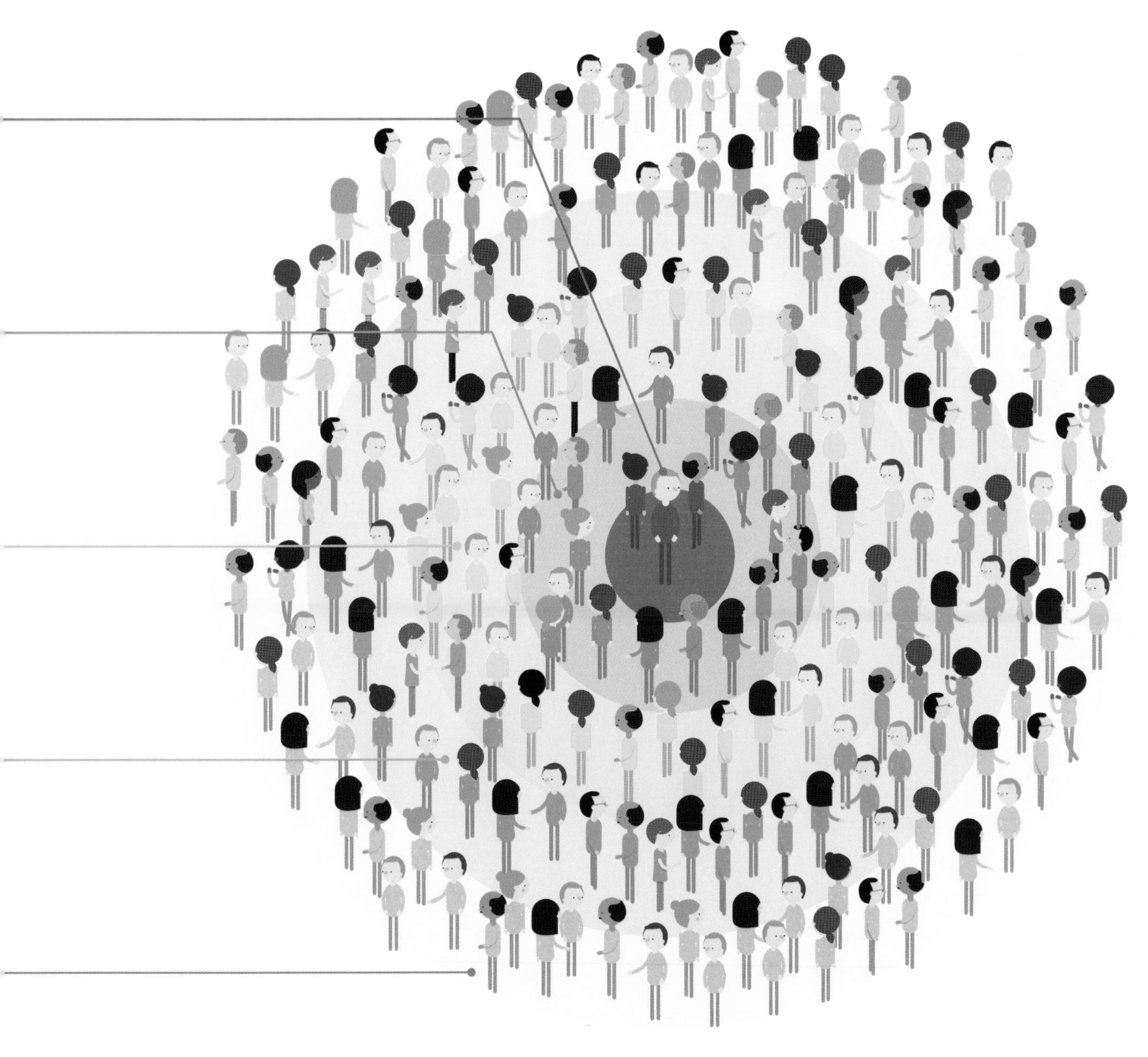

Steve Blank, Bob Dorf, *The Start-up Owner's Manual*, 2012.

120

2.4

Making Choices

ASSESSMENT
TO
TESTING
7 QUESTIONS
FINANCE
BUSINESS MODEL
FAILED

10 Questions to Assess Your Value Proposition

OBJECTIVE
Unearth potential to improve your value proposition

OUTCOME
Value proposition assessment

Use the 10 questions of great value propositions we presented previously to constantly assess the design of your value propositions. Draw on them to integrate your customer insights. Integrate them when you decide which prototypes to explore further and test with customers.

Do this exercise online

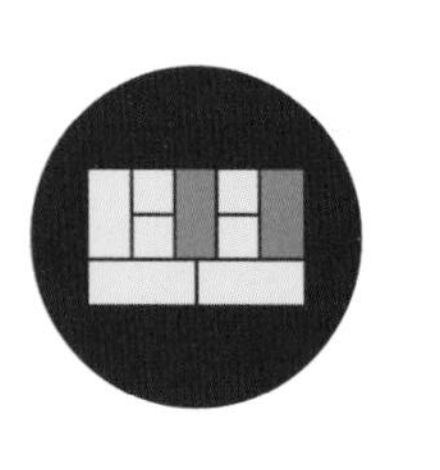

1

Is it embedded in a great business model?

2

Does it focus on the most important jobs, most extreme pains, and most essential gains?

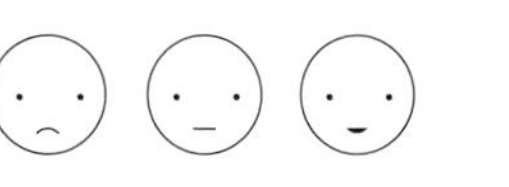

3

Does it focus on unsatisfied jobs, unresolved pains, and unrealized gains?

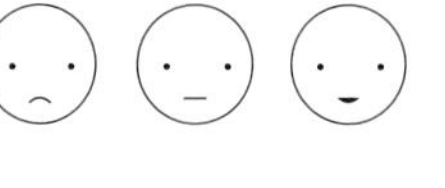

4

Does it concentrate on only a few pain relievers and gain creators but does those extremely well?

5

Does it address functional, emotional, and social jobs all together?

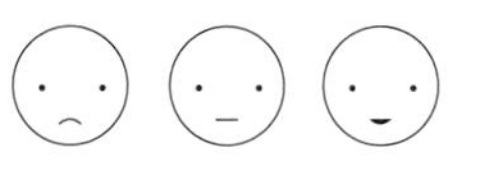

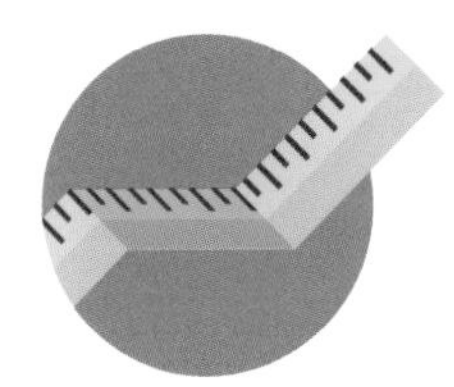

6
Does it align with how customers measure success?

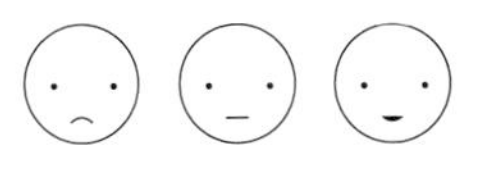

7
Does it focus on jobs, pains, or gains that a large number of customers have or for which a small number are willing to pay a lot of money?

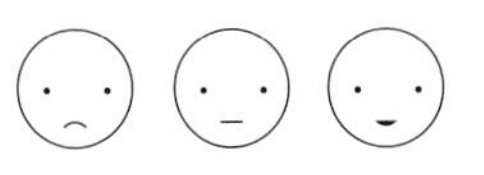

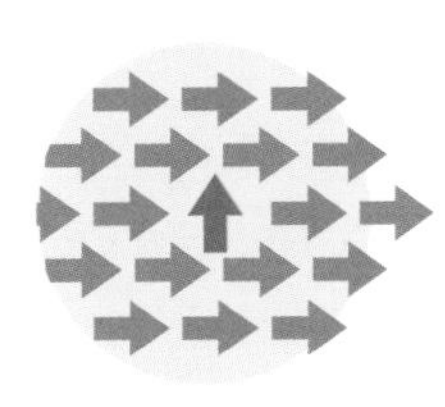

8
Does it differentiate from competition in a meaningful way?

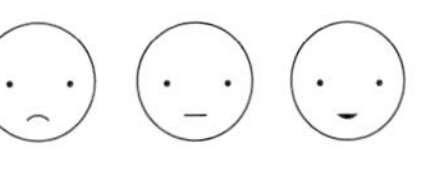

9
Does it outperform competition substantially on at least one dimension?

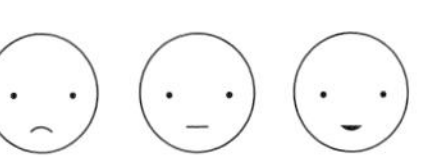

10
Is it difficult to copy?

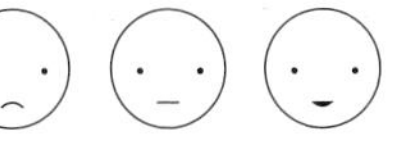

Simulate the Voice of the Customer

OBJECTIVE	OUTCOME
Stress-test your value proposition "in the meeting room"	More robust value proposition before validating it "in the market"

Use role-playing to bring the voice of the customer and other stakeholder perspectives "into the room" long before you test your value propositions in the real world.

The success of your value proposition typically depends on a number of key stakeholders. Customers are the obvious one, but there are many others (e.g., stakeholders inside your company). Pick the most important ones, and organize role-plays to stress test your value proposition from the perspective of these stakeholders.

Tips

- Make sure you choose the person who plays the stakeholder wisely. Who best represents the voice of the customer? Is it sales, customer support, field engineering, or somebody else who is close to the buyer?
- Role-plays don't replace testing your value propositions in the real world with customers and stakeholders, but they help evolve your ideas by taking a stakeholder's perspective into consideration.
- Role-plays can be an effective way to bring in the voice of the customer after you intensively analyzed customer behavior.

Two workshop participants engage in a role-playing game in which one person plays a company sales rep and the other, a stakeholder, for example, the customer. A third person takes notes.

The salesperson

Quickly evaluate your ideas with role-playing by simulating the voices of key players.

Customers

Take the customers' point of view and focus on customer jobs, pains, and gains and competing value propositions. In a B2B context think of end users, influencers, economic buyers, decision makers, and saboteurs.

Chief executive officer (CEO), senior leaders, and board members

Take the company leadership's point of view (e.g., CEO, chief financial officer [CFO], chief operations officer [COO]). Give feedback from the perspective of the company's vision, direction, and strategy.

Other internal stakeholders

Who else's buy-in in the company do you need for your idea to succeed? Does production play a role? Do you need to convince sales or marketing?

Strategic partners

Your value proposition may rely on the collaboration with strategic partners. Are you offering them value?

Government officials

What role does the government play? Is it an enabler or a barrier?

Investors/shareholders

Will they support or resist your ideas?

Local community

Are they affected by your ideas?

The planet!

What effect does your value proposition have on the environment?

Understand the Context

Value propositions and business models are always designed in a context. Zoom out from your models to map the environment in which you are designing and making choices about the prototypes to pursue. The environment is made up of competition, technology change, legal constraints, changing customer desires, and other elements. Learn more with the illustration on this spread or read more in *Business Model Generation*.

Zoom out

Industry Forces

Key actors in your space, such as competitors, value chain actors, technology providers, and more

Macroeconomic Forces

Macro trends, such as global market conditions, access to resources, commodities prices, and more

Key Trends

Key trends shaping your space, such as technology innovations, regulatory constraints, social trends, and more

Market Forces

Key customer issues in your space, such as growing segments; customer switching costs; changing jobs, pains, and gains; and more

Zoom in

Osterwalder & Pigneur, Business Model Generation, 2010.

Participatory TV

Imagine you are a player in the movie industry. So far, you have been making movies and TV series with leading actors for cinema and home viewers globally. But you'd like to explore new avenues.

There is one idea that your innovation teams want to explore more closely: participatory TV—enabling viewers to crowdsource the plot of a TV series.

Illustration: Participatory TV

Sketch out your environment and ask which elements look like...

- an opportunity that strengthens the case for your value proposition (in **green**)
- a threat or a constraint that undermines or limits it (in **red**)

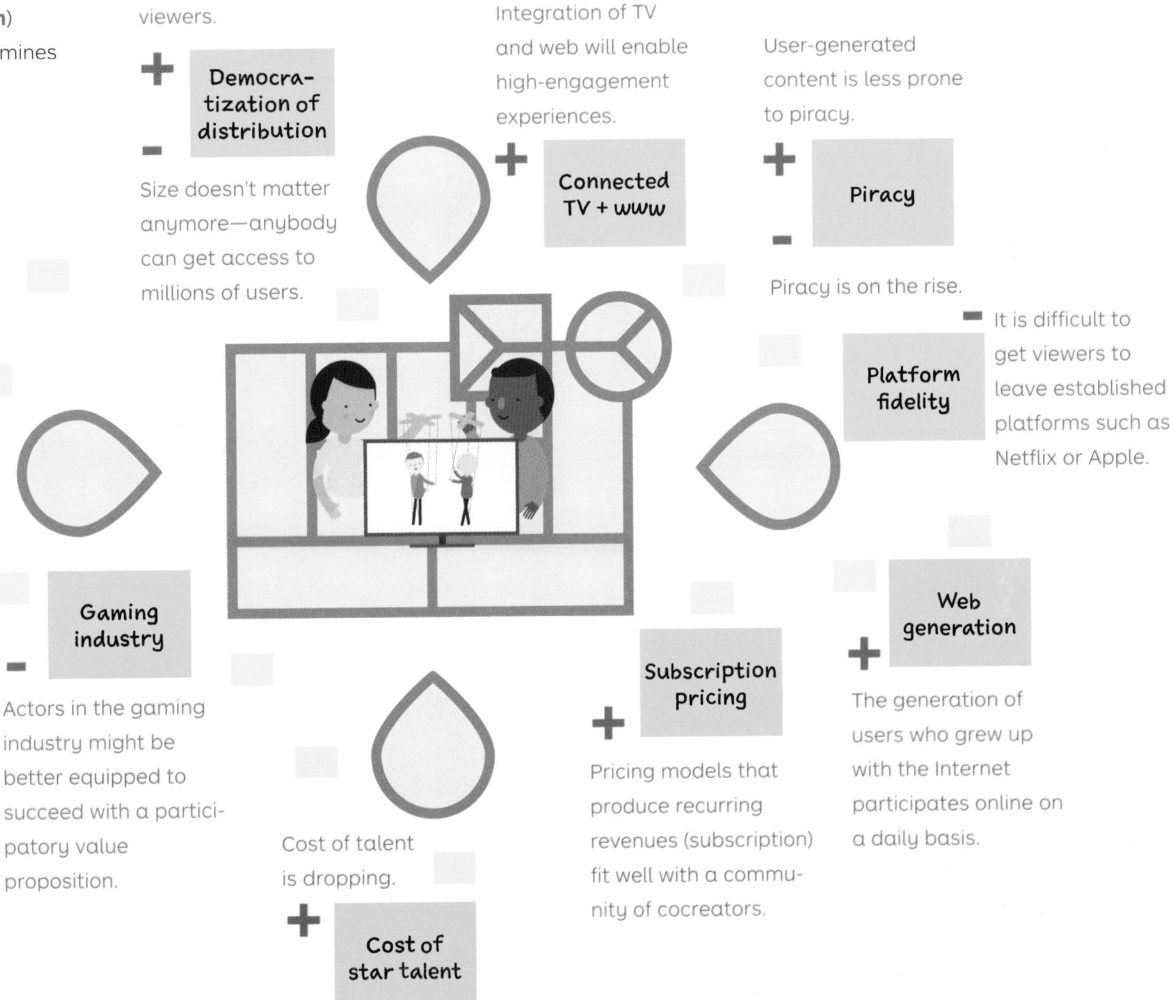

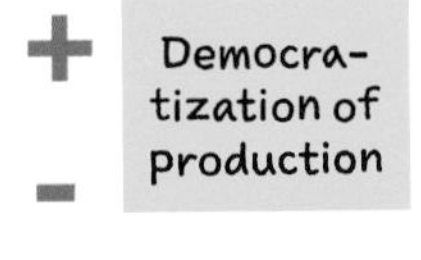

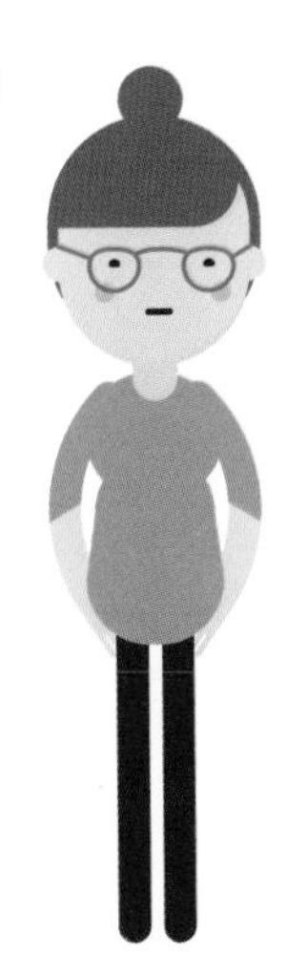

Value Proposition Design vs. Competitors

Let's focus on one element of your design and decision making environment: your competitors. Assess how your value proposition performs against those of your competition by comparing them on a Strategy Canvas, a graphical tool from the *Blue Ocean Strategy* book. This is a simple but powerful way to visualize and compare how the "benefits" of your value proposition perform.

On this spread we compare the performance of *Value Proposition Design* to the performance of executive education and massive open online courses (so-called MOOCs). We do so by drawing a Strategy Canvas with a number of competitive factors on the *x*-axis and then plot how the different competitors perform on each one of these factors. We selected the competitive factors from our value map and complemented them with elements from our competitors' value maps.

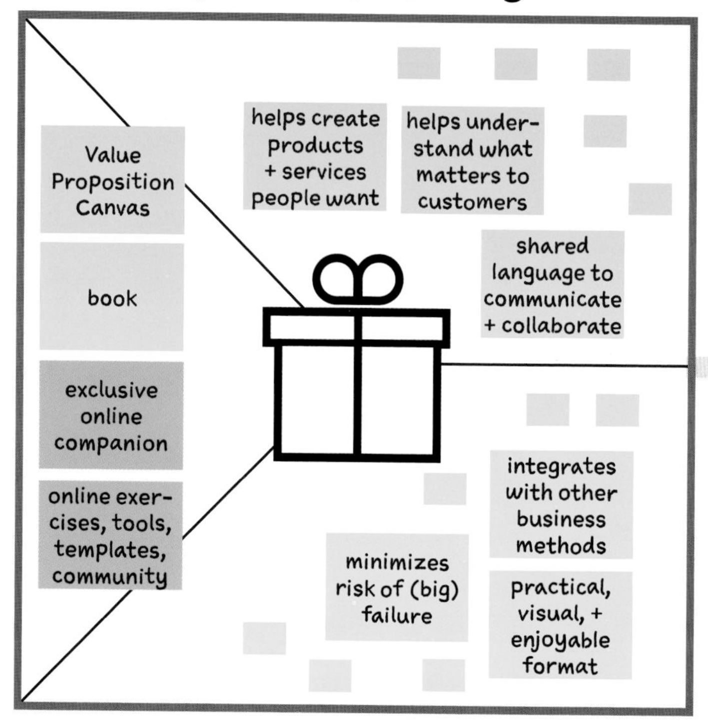

Select the most important features from your VP to use on the Strategy Canvas as factors of competition.

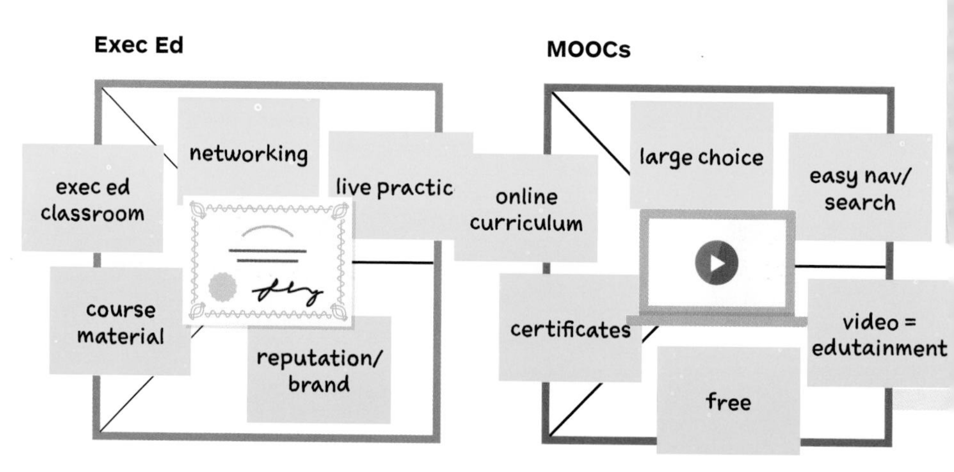

Kim & Mauborgne, *Blue Ocean Strategy*, 2005.

Strategy Canvas

Value proposition of this book vs. executive education vs. MOOCs as compared with each other

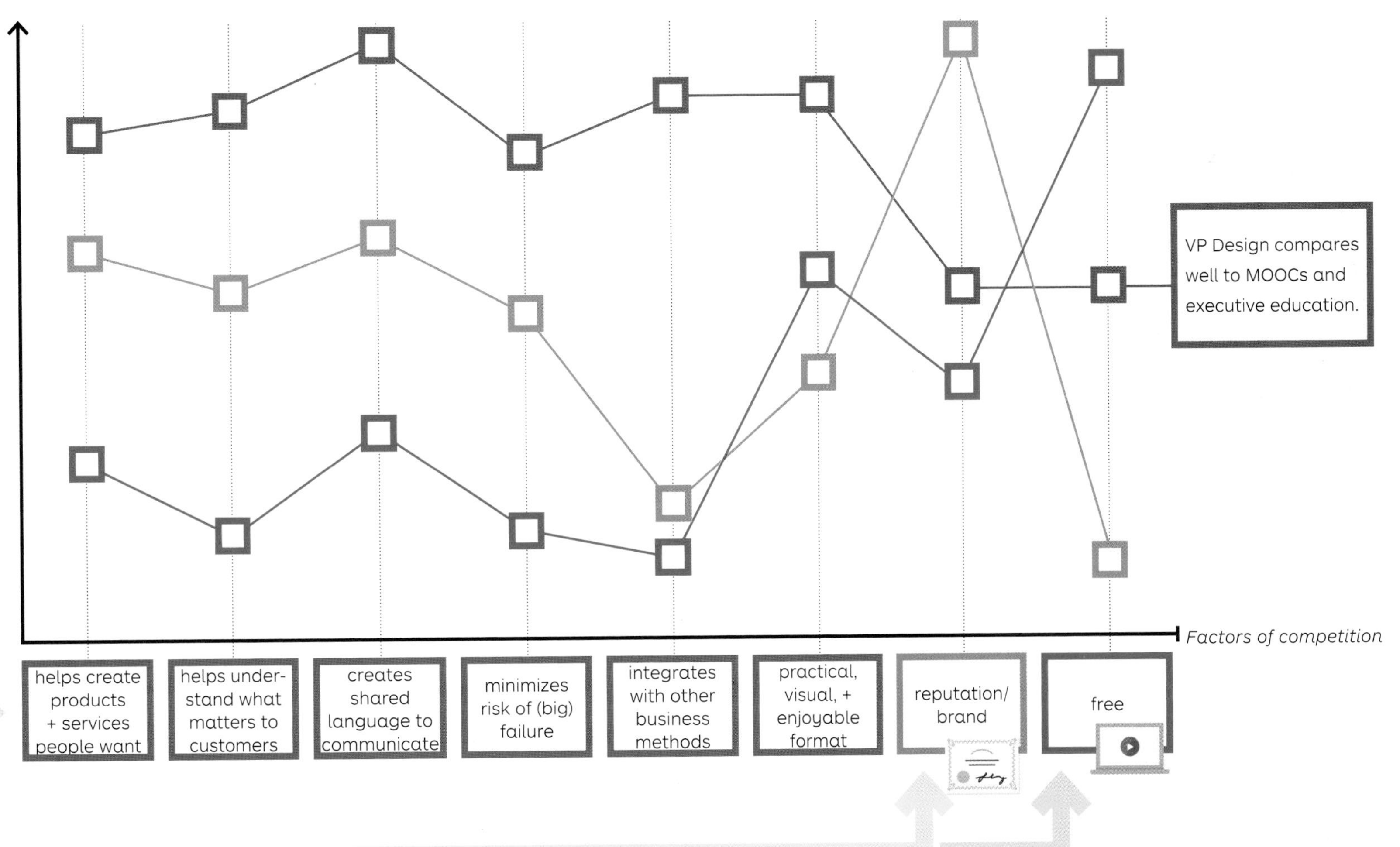

Compare Your Value Proposition with Competitors

OBJECTIVE	OUTCOME
Understand how you are performing compared to others	Visual comparison with competitors

Use the Strategy Canvas from the *Blue Ocean Strategy* book to plot the performance of your value proposition against those of your competitors. Then compare the curves to assess how you are differentiating.

Instructions

Draw a Strategy Canvas step by step and compare your value proposition with those of your competitors.

1. Prepare or pick a value map for this exercise.
2. Grab a big sheet of paper or use a whiteboard.
3. Follow the steps.

1

Select a value proposition.

Select the value proposition (prototype) you want to compare.

2

Select factors of competition.

Draw a horizontal axis (x-axis). Pick the pain relievers and gain creators you want to compare with competition. Place them on the axis. These are the factors of competition of your Strategy Canvas.

Tip

You can also add pains and gains if you feel like they better describe important factors of competition.

3

Score your value proposition.

Draw a vertical axis (y-axis) to represent the performance of a value proposition. Add a scale from low to high or from 0 to 10. Plot how your value proposition performs on each factor of competition on the x-axis (i.e., the pain relievers and gain creators you chose).

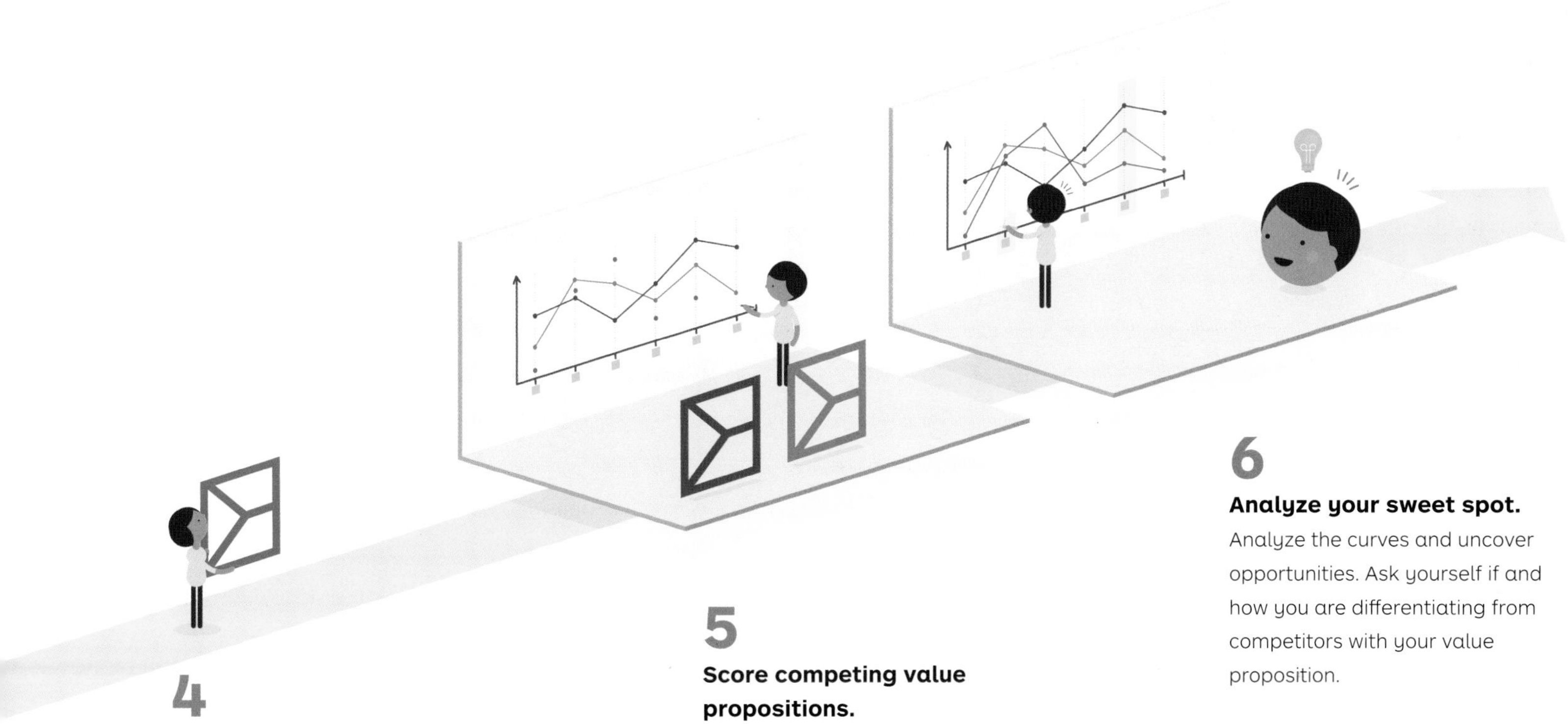

4

Add competing value propositions.

Add competing value propositions to the Strategy Canvas. Choose those that are most representative of the competition out there. Add pain relievers and gain creators from their value proposition to the factors of competition on the *x*-axis if necessary.

Tip

Consider competing value propositions beyond traditional industry boundaries. Don't just compare value propositions based on products and services that are similar to yours.

5

Score competing value propositions.

Plot how competing value propositions perform, just as you did for your own.

Tip

Use this tool to compare the performance of alternative value propositions you might be considering.

6

Analyze your sweet spot.

Analyze the curves and uncover opportunities. Ask yourself if and how you are differentiating from competitors with your value proposition.

Tip

Make sure the factors of competition that you compare align with the top jobs, pains, and gains in the customer profile. Normally that should be the case, because pain relievers and gain creators are designed to match relevant jobs, pains, and gains.

Avoid Cognitive Murder to Get Better Feedback

Present your value proposition to others to gather feedback, get buy-in, and complement the more "analytical assessment" that we looked at up to this point and the experiments we will study in the testing chapter.

Make sure you get the best from presenting your ideas by explaining them with disarming simplicity and coherence. It would be a waste of time and resources to put all your energy into designing remarkable value propositions only to fail to present them in a convincing way when it matters.

Presenting your ideas and canvases in a clear and tangible way is critical throughout the design process. Present early and rough prototypes before refining to get buy-in from different stakeholders. Only work on more refined presentations later in the design process.

One of the most important aspects of presenting value propositions is to convey messages with customer jobs, pains, and gains in mind. Never just pitch features; instead, think about how your value proposition helps get important jobs done, kills extreme pains, and creates essential gains.

Use low-fidelity prototypes to make your ideas tangible.

Always refer back to customer jobs, pains, and gains in your presentation.

Best Practices for Presenters

√ DOS	× DON'TS
Simple	Complex
Tangible	Abstract
Presenting only what matters	Presenting all you know
Customer-centric	Feature-centric
1 piece of info after the other	All information at once
The right media support	No visual support
Storyline	Random flow of information

1. Start with an empty canvas. Make sure listeners were given at least a short introduction to the canvas.

2. Begin your presentation wherever it makes most sense. You can start with products or with jobs.

3. Put up one sticky note after the other progressively to explain your value proposition so your audience doesn't experience cognitive murder. Synchronize what you say and what you put up. Tell a story of value creation by connecting products and services with customer jobs, pains, and gains.

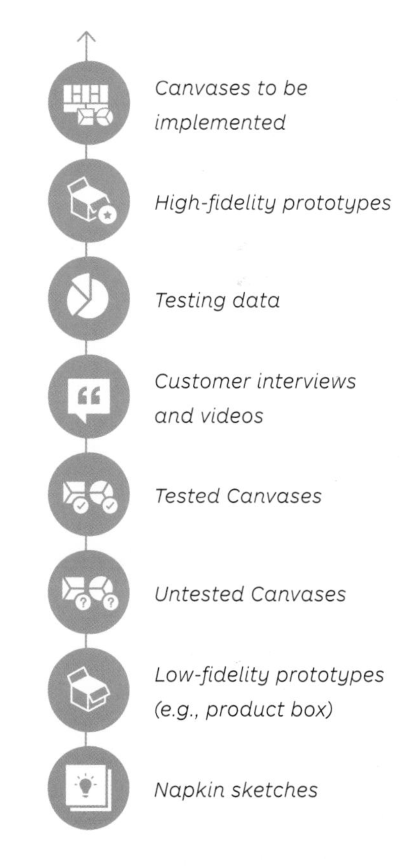

What to present and when
Present different types of prototypes depending on how far you are into the design and testing process.

Master the Art of Critique

Practice the art of feedback to help ideas evolve rather than stall. This goes for feedback receivers who present ideas, as well as for feedback providers who give input on ideas.

Learn from the design professions, where people are trained to present ideas early and feedback providers are trained to provide effective design critiques. This contrasts with feedback providers in business who are often leaders in steering boards or advisory committees. They are trained to decide rather than to give feedback. If they cannot get to decisions fast, they often get nervous or become unsatisfied.

Teach feedback providers how to help ideas evolve (rather than to decide on them). Get them to understand that value proposition prototypes are still rough and evolving during the design and testing phase. Prototypes may radically change, in particular based on market facts that matter more than the opinion of feedback providers.

Teach feedback receivers that feedback providers are not as important as customers, however powerful they might be. Listening to feedback providers more than to customers and market facts only postpones failure.

Get "Master the Art of Critique" poster

In a great feedback culture...

People feel comfortable presenting (bold) new ideas early, knowing that they will evolve substantially, maybe into something very different.

Present early.

Distinguish between Three Types of Feedback

		+	-
OPINION	*"If we added ___ I believe we'd have a better chance to make it work."*	Logical reasoning can help improve ideas.	It can lead to pursuing pet ideas of people with more power.
EXPERIENCE	*"When we did ___ in our last project, we learned that..."*	Past experiences provide valuable learning that can help prevent costly mistakes.	Failing to realize that different contexts lead to different results.
(MARKET) FACTS	*"We interviewed people about this and learned that ___ percent struggled with..."*	This provides input that reduces uncertainty and (market) risk.	Measuring the wrong data or simply bad data can lead to missing out on a big opportunity.

Leaders and decision makers are trained to give feedback on early ideas to help them evolve. They know their opinion can be trumped by market facts and they're comfortable with that.

Don't ✗	Shoot people down for presenting new (bold) ideas.	Present only refined ideas to leadership and decision makers.	Have long, unstructured, free-flowing, time-consuming discussions.	Allow for proliferation of pure opinion.	Create a context that enables politics and personal agendas to supersede value creation.	Create negative vibes that destroy positive creative energy.	Foster culture in which feedback destroys big ideas because they're hard to implement.	Just ask "why?"
Do ✓	Create a safe environment in which people feel comfortable to present (bold) ideas.	Foster a culture of early feedback on rapidly evolving ideas.	Run facilitated, structured feedback processes.	Provide feedback based on experience or (market) facts.	Encourage a customer-centered feedback culture that neutralizes politics.	Bring in fun and productive feedback processes.	Draw a distinction between hard to do and worth doing.	Ask, "Why not?" "What if?" and "What else?"

Collect Efficient Feedback with de Bono's Thinking Hats

Collect feedback on ideas, value propositions, and business models using Edward de Bono's thinking hats. This method is very effective—especially in large groups—and helps you avoid losing time in endless discussions.

OBJECTIVE
Collect feedback effectively and avoid lengthy discussions

OUTCOME
Understanding of what's good or bad about ideas and how they can be improved

Workshop participants put on a metaphorical colored hat that symbolizes a certain type of thinking. This technique allows you to quickly collect different types of feedback and avoid having an idea shot down for purely political reasons. Use four of de Bono's six thinking hats to gather feedback.

1

Pitch

3–15 min depending on stage of idea

The design team presents their idea and value proposition and/or Business Model Canvas.

2

White hat

Information and data; neutral and objective

2–5 min depending on stage of idea

"Audience" members ask clarifying questions to fully understand the idea.

3a

Black hat

Difficulties, weaknesses, dangers; spotting the risks

1 min to write down

Participants write down why it's a bad idea on a sticky note.

3 min to collect feedback

The facilitator rapidly collects one feedback after the other on a flip chart, while participants read it out loud.

Tips

- This exercise requires strong facilitation skills. Make sure people don't voice opinions when it's time for the white hat to ask clarifying questions.
- Make sure that regardless of whether people hate or love an idea, everybody puts on all hats, white, black, yellow, and green.
- Use the black hat before the yellow hat to neutralize extremely negative people. Once they voiced their feedback, they might even think positively.
- De Bono's Thinking Hats also works well in small groups or individually to help people come up with all the reasons why an idea might fail or succeed.

4a

Yellow hat

Positives, plus points; why an idea is useful

1 min to write down
Participants write down why it's a good idea on a sticky note.

4b

3 min to collect feedback
The facilitator rapidly collects one feedback after the other on a flip chart, while partici-pants read it out loud.

5

Green hat

Ideas, alternative, possibilities; solutions to black hat problems

5–15 min of open discussion
The floor is opened to discussion. Participants bring in suggestions regarding how to evolve the ideas that were presented.

6

Evolve

The presenting team evolves their idea equipped with the white, black, yellow, and green hat feedback.

Edward de Bono, Six Thinking Hats, 1985.

Vote Visually with Dotmocracy

OBJECTIVE
Visualize the preferences of a group and avoid lengthy discussions

OUTCOME
Quick selection of ideas

Use Dotmocracy to quickly visualize the preferences of a group, in particular in large workshop settings. This is a simple and speedy technique to prioritize among different value propositions and business model options. It helps to prevent lengthy discussions.

1

Idea gallery.
Ideas or canvases are exposed on a wall as a gallery of options.

2

Stickers.
Each workshop participant gets the same number of stickers (e.g., 10), and each sticker counts as one vote.

3

Criteria.
The voting criteria are defined. For example, participants may be instructed to place a sticker on their favorite ideas.

4

Vote.
Participants can put all their stickers on one idea or distribute them across several ideas.

5

Count.
Stickers are counted, and preferred ideas are highlighted.

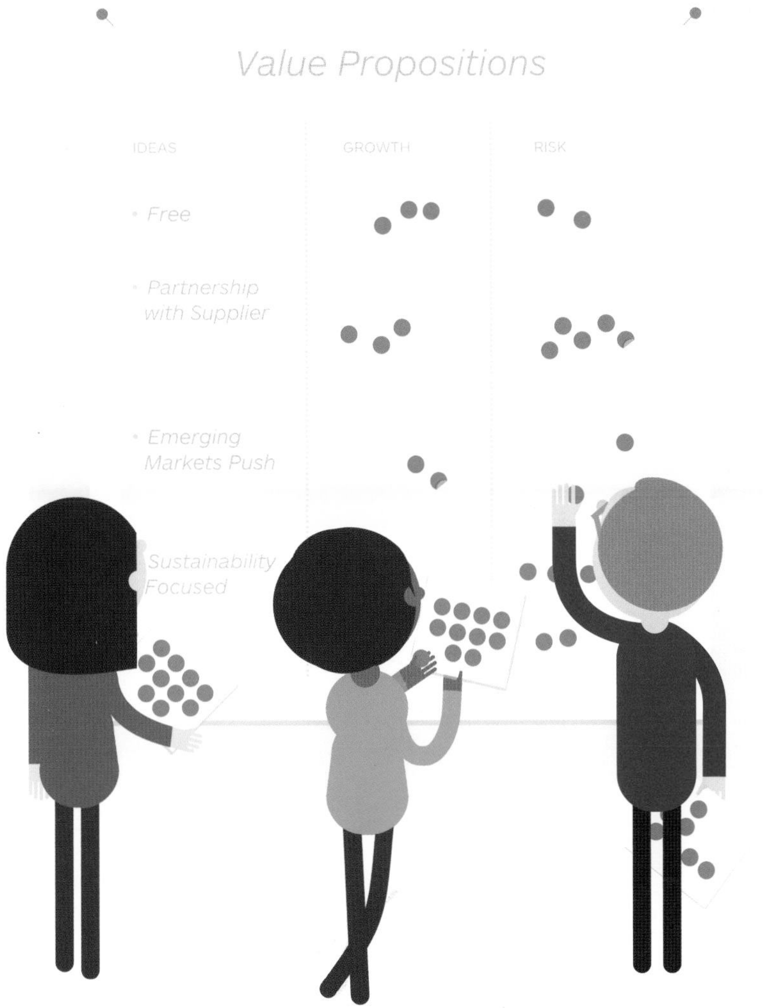

Multicriteria

Use a table when you want to use several criteria to select among alternative value propositions and business models.

Dotmocracy is used to select ideas based on internal criteria, such as growth potential, risk, and differentiation potential. Apply this technique during the design process to choose among several alternatives before you test them in the real world.

Define Criteria and Select Prototypes

OBJECTIVE	OUTCOME
Select among a range of alternatives	Ranking of prototypes

Decide which criteria are most important to you and your organization and select value propositions and business models accordingly during the design process. You need to prioritize among (hopefully attractive) alternatives, even if your customer is the final judge of your ideas later on in the process.

1

Brainstorm criteria.

Come up with as many criteria as you can to assess the attractiveness of your prototypes.

Use the following themes and criteria as an input for your own selection criteria.

Fit with Strategy

How the idea fits with the overall direction of the company

- Aligns with strategy
- Good timing
- Fits with desired risk level
- Can replace outdated business models

Fit with Customer Insights

How the idea relates to the first customer insights gained during first market research

- Important job
- No good solution exists
- Visible and tangible pain
- Strong customer evidence

Competition and Environment

How the idea allows the company to position itself related to the competition

- Provides competitive advantage
- Fits with tech and other trends
- Allows differentiation

Relation to Current Business Model

How the idea builds or doesn't build on the current business model

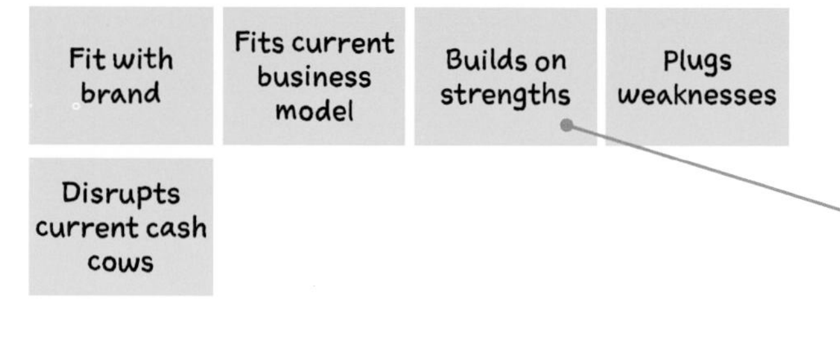

Financials and Growth

What potential each idea has related to growth and financials

- Market size
- Revenue potential
- Market growth
- Margins

Implementation Criteria

How difficult it is to implement the idea from design to market

- Time to market
- Cost to build
- Do we have right team and skills
- Access to target customers
- Technology risk
- Implementation risk
- Risk of management resistance

2
Select criteria.
Select the criteria that are most important for your team and organization.

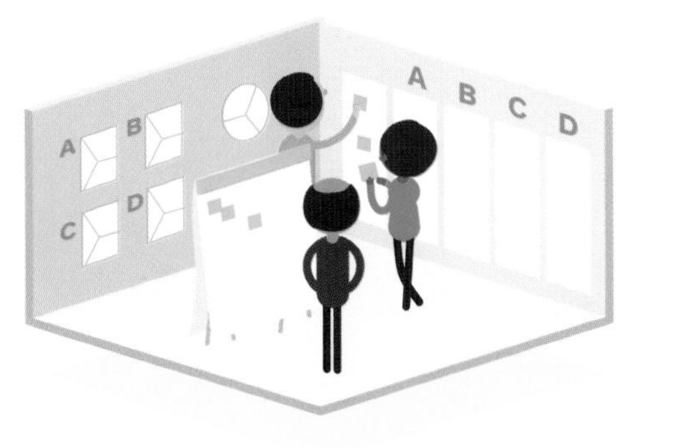

Criteria	Prototype A: 36	Prototype B: 32	Prototype C: 12	Prototype D: 42
Allows differentiation				
Builds on strengths				
Market growth				

3
Score prototypes (0 [low] – 10 [high]).
Score each idea on the criteria you chose.

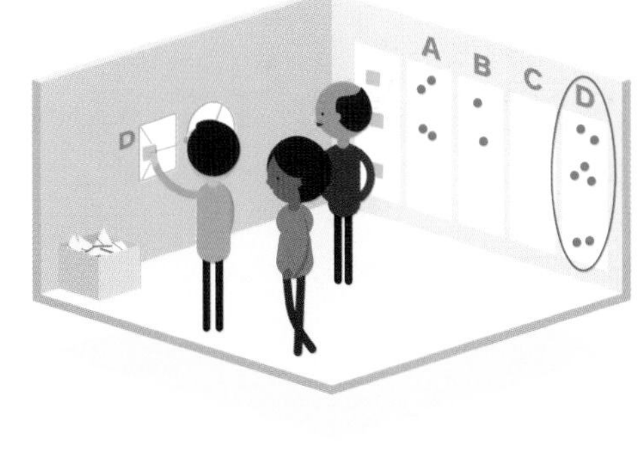

4
Evolve prototype and explore with market.
Evolve your prototype (e.g., based on the scores it got), and test it in the market to learn if it really has potential.

2.5

Finding the Right Business Model

ASSESSMENT
TO TESTING
7 QUESTIONS
FINANCE
BUSINESS MODEL
PROTOTYPING
FAILED

Create Value for Your Customer *and* Your Business

To create value for your business, you need to create value for your customer.

To sustainably create value for your customer, you need to create value for your business.

A business that generates fewer revenues than it incurs costs will inevitably disappear, even with the most successful value proposition. This section shows how getting both the business model and the value proposition right is a process of back and forth until you nail it.

Are you creating value for your business?
The Business Model Canvas makes explicit how you are creating and capturing value for your business.

Zoom out to the bigger picture to analyze if you can profitably create, deliver, and capture value around this particular customer value proposition.

-Zoom

+Zoom

Zoom in to the detailed picture to investigate if the customer value proposition in your business model really creates value for your customer.

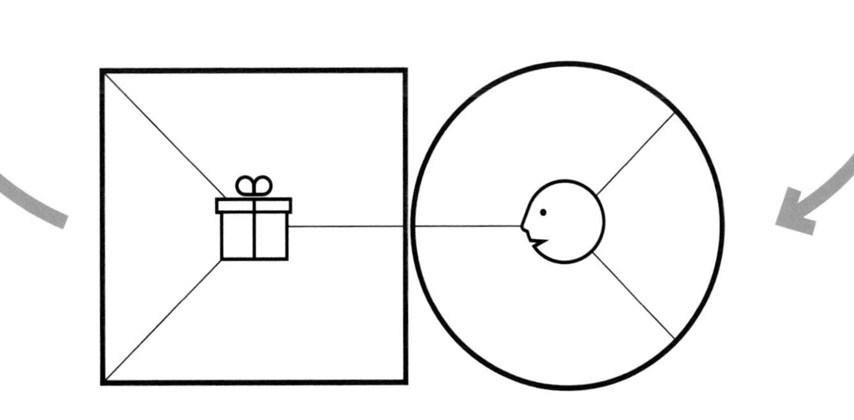

Are you creating value for your customer?
The Value Proposition Canvas makes explicit how you are creating value for your customers.

Azuri (Eight19): Turning a Solar Technology into a Viable Business

1.6 billion people in the world still live without electricity. Could innovative value propositions and business models around new technology offer answers?

Simon Bransfield-Garth founded Eight19 based on a printed plastic technology originating from Cambridge University. The technology is designed to deliver low-cost solar cells. In 2012 Eight19 launched Azuri to commercialize the technology and bring electricity to off-grid customers in rural emerging markets.

Finding the right value propositions and business models in such a context is not easy. We illustrate how it is a continuous back and forth between both on the following pages.

Azuri Business Model: version 0

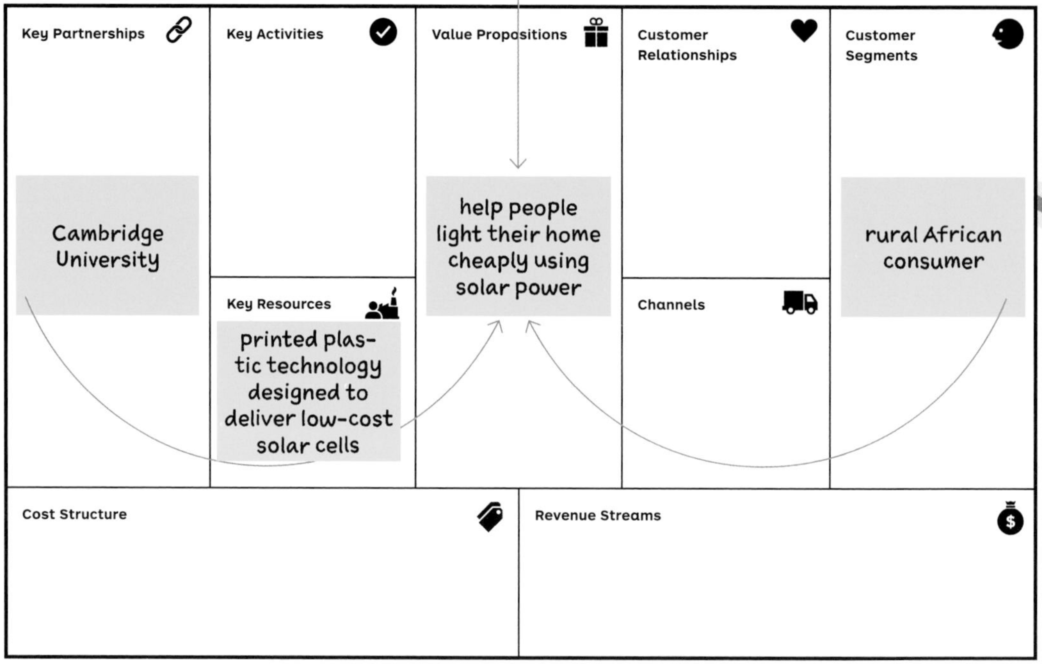

1

Initial idea

An opportunity.

Developing low-cost solar technology and providing low-income people with access to electricity.

Case adopted in accordance with Azuri.

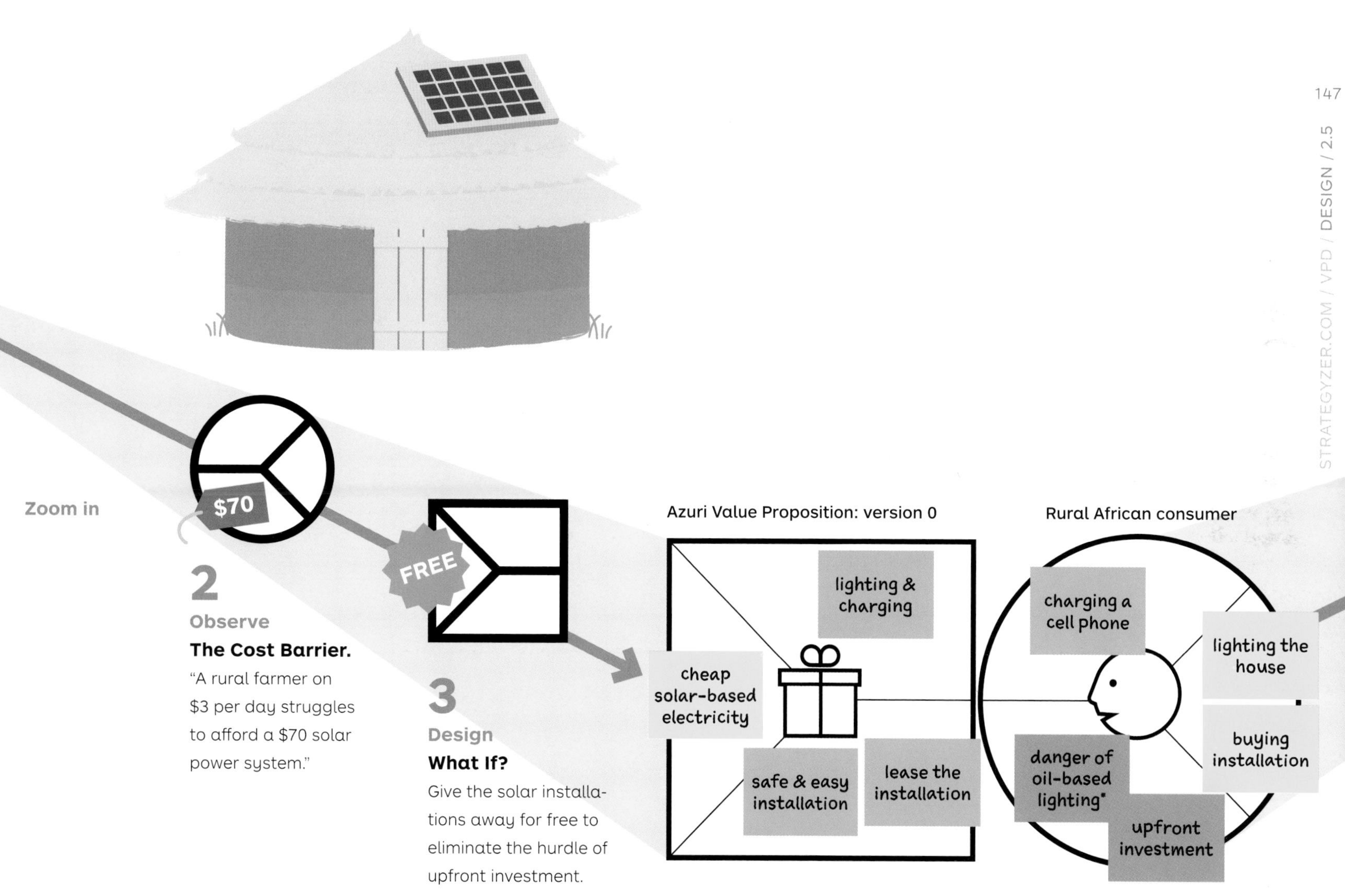

2
Observe
The Cost Barrier.
"A rural farmer on $3 per day struggles to afford a $70 solar power system."

3
Design
What If?
Give the solar installations away for free to eliminate the hurdle of upfront investment.

An alternative for lighting is burning oil, which is dangerous and expensive.

4

Iteration 2

Idea for Business Model.

Lease the solar installations and collect regular subscription fees; it works just fine with conventional panels; get resources and partnerships for financing the installations.

Zoom out

Azuri Business Model: version 1

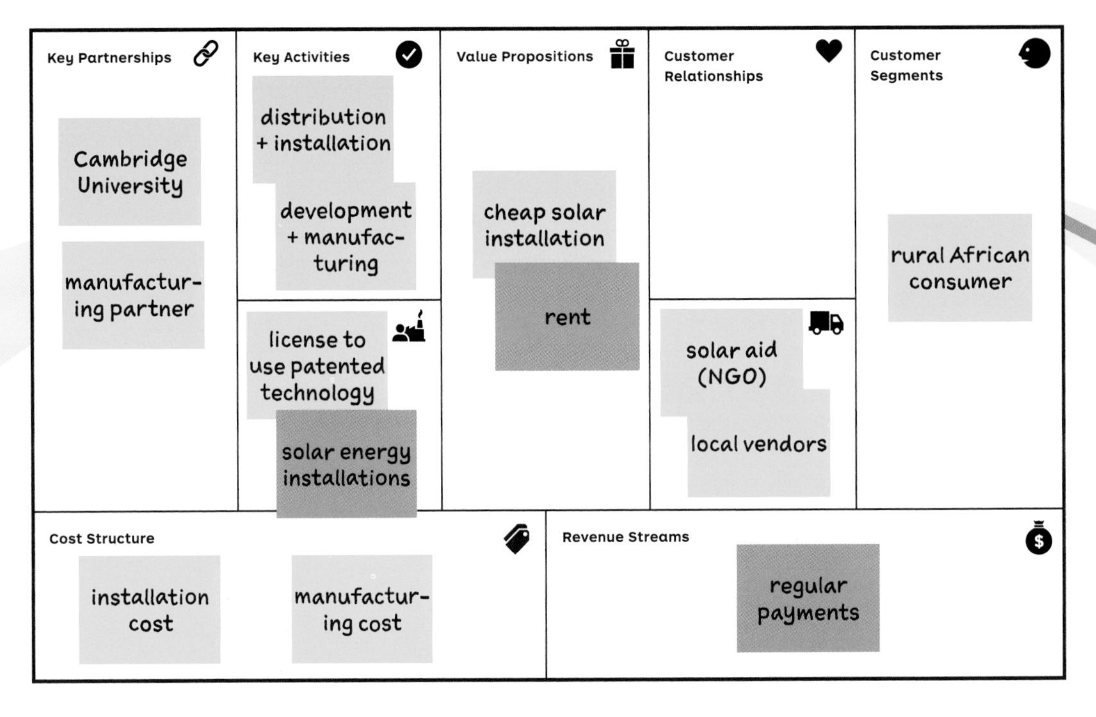

So...

How does the *Indigo* value proposition look for a customer?

$10 — $1 — free — upgrade — Time

$10: **Buy the Indigo kit (solar panel, lamps, charger).**

$1: **Buy scratch cards, use SMS from a mobile phone, enter the resulting passcode into the *Indigo* unit, and use the installation for a period of time (typically a week).**

free: **Own your box after 80 scratch cards, or...**

upgrade: **Escalate to a larger system and access more energy; continue to buy scratch cards.**

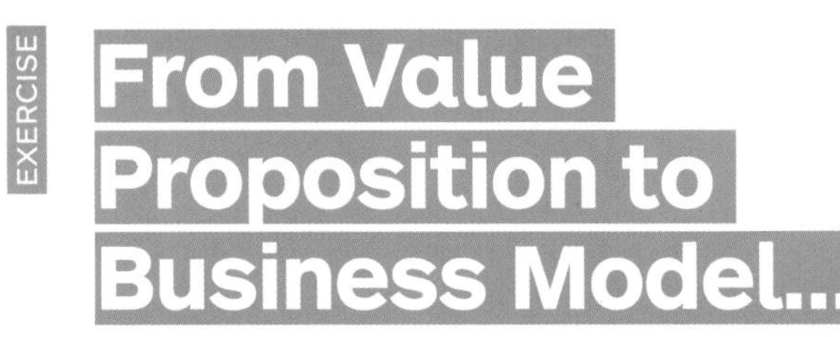

From Value Proposition to Business Model...

OBJECTIVE
Practice the connection between value proposition and business model with no risk

OUTCOME
Improved skills

From p. 96

A1
Front Stage.
Prototype a revenue model, select distribution channels, and define the relationships that could be adopted with customers.

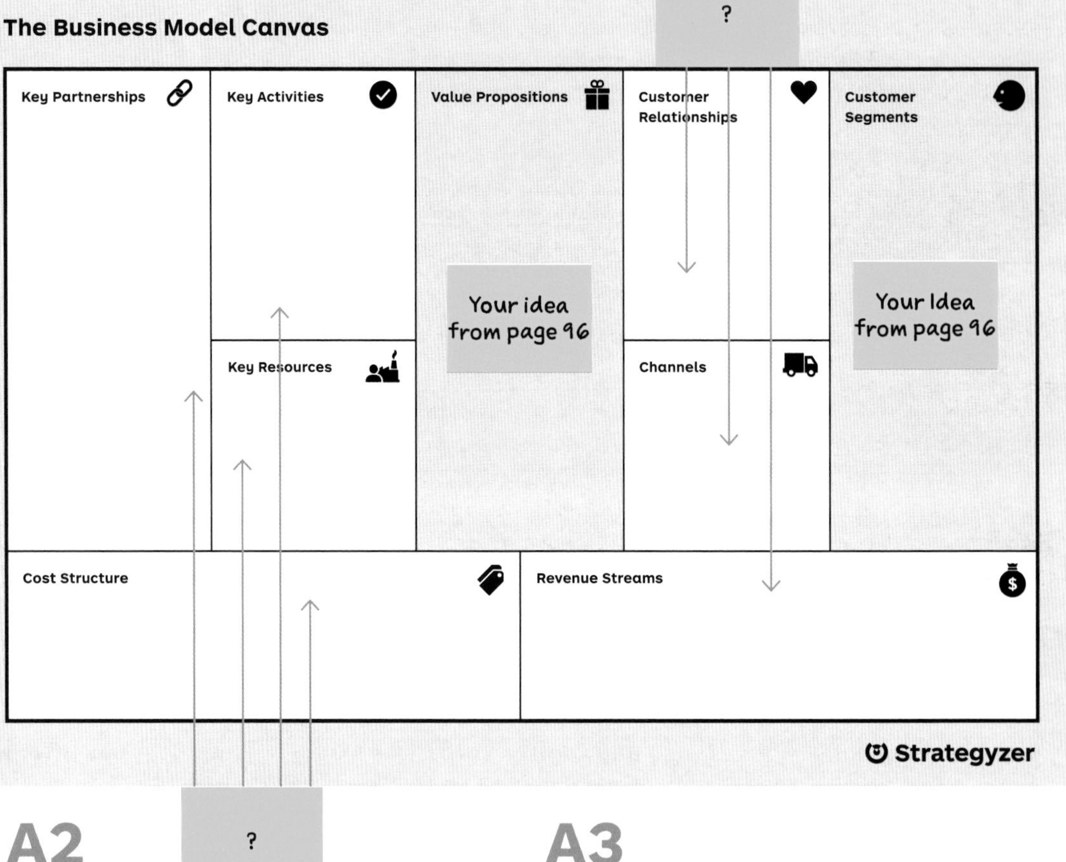

Part A
Design the Full Business Model.
On page 96 you imagined a value proposition to commercialize an innovative compressed air energy storage technology. Now map out the remaining business model elements and sketch out the rough numbers (part A).

A2
Backstage.
Add the Key Resources, Key Activities, and Partners required for the model to work and use that to estimate the cost structure.

A3
Assessment.
Assess your prototype and detect possible weaknesses of the business model ➔ p. 156.

...and Back Again

Part B

Revisiting the Value Proposition
Assess the weaknesses of your first full business model prototype (from part A). Ask yourself how you could improve or change your initial value proposition, maybe by shifting to an entirely different segment by considering the following five questions:

Zoom in

B1
New VP?
Could there be another radically different value proposition for the same technology?

B2
New segment?
Will you keep the same customer segment, or will you shift to an entirely different, maybe larger, market segment?

B3
Refine or clear your profile?
Could you refine your customer profile, or do you need to describe an entirely new one because you switched customer segments?

B4
Change or clear your benefits?
Do you need to change or clear the benefits your value proposition created because the customer profile changed?

B5
Got fit?
Do you have fit between your new customer profile and the newly designed value proposition?
➔ p. 40 on fit

Repeat Step A if required.

The Value Proposition Canvas

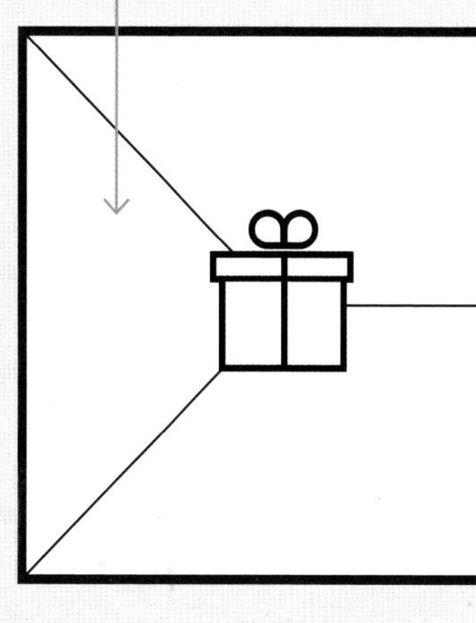

Tip
Follow up on your new customer assumptions by researching customers ➔ p. 106 and producing evidence ➔ p. 216.

Stress Testing with Numbers: A MedTech Illustration

A great value proposition without a financially sound business model is not going to get you very far. In the worst case you will fail because your business model incurs more costs than it produces revenues. But even business models that work can produce substantially different results.

Play with different business models and financial assumptions to find the best one. We illustrate this with the medical technology illustration on this spread. We sketched out two models both starting from the same technology that enables building a cheap diagnostic device.

Prototype 1 generates $5.5 million in revenues and a profit of $0.5 million. Prototype 2 starts from the same technology but produces more than $30 million in revenues and a profit of $23 million with a different value proposition and business model.

Only the market can judge if either model could work, but you certainly want to explore and test the best options.

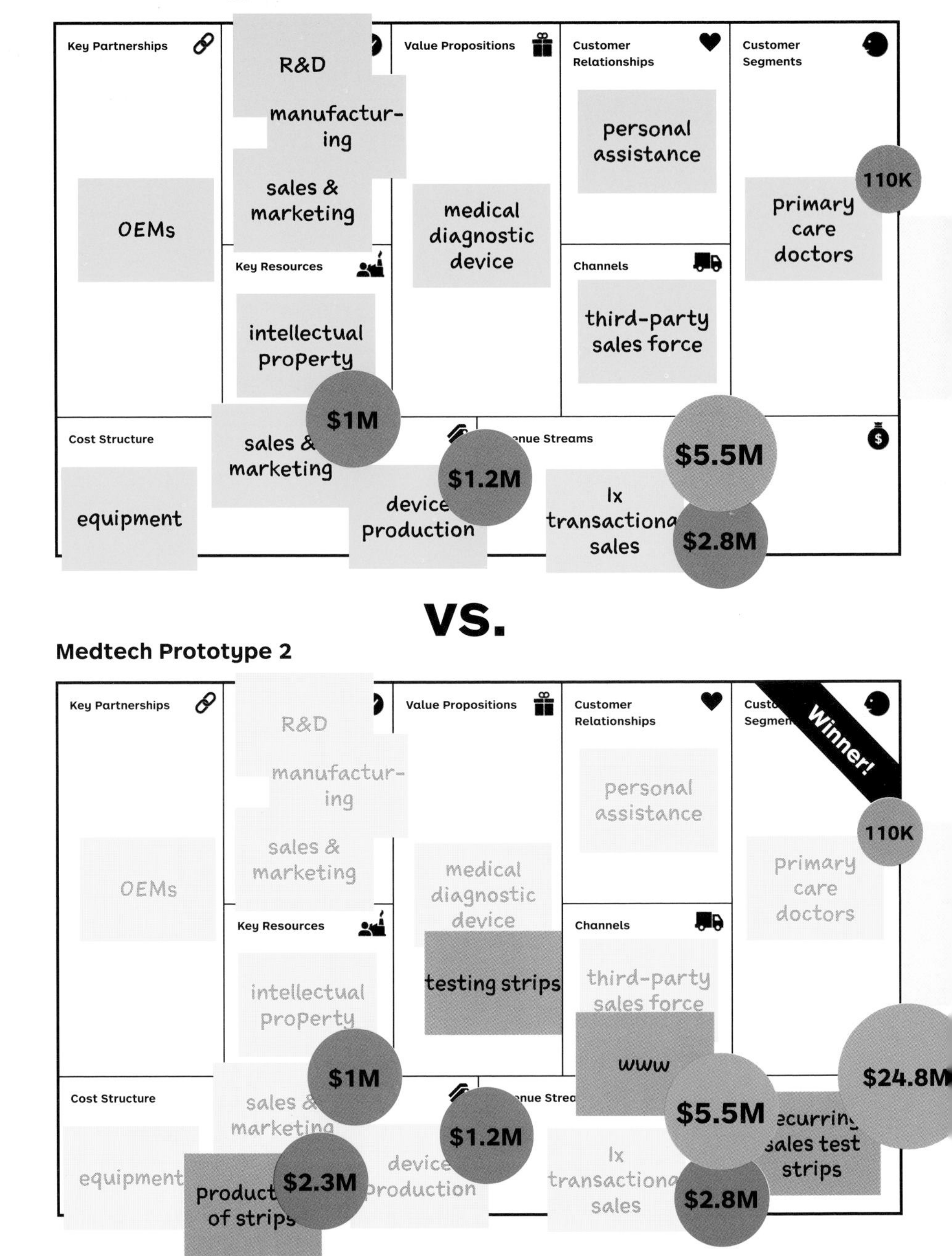

Model 1: Sales of Medical Diagnostic Device

- 1× transactional sales of device to primary care doctors in the United States for $1,000/device
- 5 percent market share
- Sales via third-party sales force -50 percent commission
- Variable production costs of $225/device
- Fixed marketing expenses of $1 million

Costs	Revenues
Device Production, 1.2M	Device Sales, 5.5M
Sale & Marketing, 1M	
Sales Commission, 2.8M	
Profit 0.5M	

$0.5M

Profit

A quick sketch of the numbers provides us with a sanity check that this model is not very profitable, so we should go back and explore changes to the business model.

Value Prop Model 1

patients don't wait for results
medical diagnostic device
immediate in-house testing

Primary Care Doctor

happy patients
no need for follow-up with patients
test patient's health risk
send to laboratory
wait for results
patients calling in for results
cleaning diagnostic devices

Model 2: Recurring Revenues from Consumable Testing Strips

- Each diagnosis requires a consumable testing strip
- Recurring revenues from selling an average of 5 strips/month/device for $75 each
- Variable production costs of testing strips of $7/strip

Costs	Revenues
Device Production, 1.2M	Device Sales, 5.5M
Sale & Marketing, 1M	Testing Strip Sales, 24.8M
Sales Commission, 2.8M	
Testing Strips Prod., 2.3M	
Profit, 23M	

Profit

The same technology with a different business model now yields a much larger potential profit. Although these numbers aren't validated, it's clearly the more interesting prototype to take to the testing stage.

Value Prop Model 2

patients don't wait for results
medical diagnostic device
testing strips
immediate in-house testing
increased hygiene via consumable strips

Seven Questions to Assess Your Business Model Design

OBJECTIVE
Unearth potential to improve your business model

OUTCOME
Business Model Assessment

Great value propositions should be embedded in great business models. Some are better than others by design and will produce better financial results, will be more difficult to copy, and will outperform competitors.

Score your business model design by answering these seven questions:

1. Switching Costs.
How easy or difficult is it for customers to switch to another company?

My customers are locked in for several years.

10
○
○
○
○
○
○
○
○
○
○
○
0

Nothing holds my customers back from leaving me.

Apple's iPod got people to copy their entire music library into the iTunes software, which made switching more difficult for customers.

2. Recurring Revenues.
Is every sale a new effort or will it result in quasi-guaranteed follow-up revenues and purchases?

100 percent of my sales lead to automatically recurring revenues.

10
○
○
○
○
○
○
○
○
○
○
○
0

100 percent of my sales are transactional.

Nespresso turned the transactional industry of selling coffee into one with recurring revenues by selling single-portioned pods that fitted only into their machines.

Download "Seven Questions to Assess Your Business Model"

3. Earnings vs. Spending.
Are you earning revenues before you are incurring costs?

I earn 100 percent of my revenues before incurring costs of goods & services sold (COGs).

10
○ ○ ○ ○ ○ ○ ○ ○ ○ ○ ○
0

I incur 100 percent of my costs of COGs before earning revenues.

Personal computers (PCs) used to be produced well ahead of selling them at the risk of inventory depreciation until Dell disrupted the industry, sold directly to consumers, and earned revenue before assembling PCs.

4. Game-changing Cost Structure.
Is your cost structure substantially different and better than those of your competitors?

My cost structure is at least 30 percent lower than my competitors.

10
○ ○ ○ ○ ○ ○ ○ ○ ○ ○ ○
0

My cost structure is at least 30 percent higher than my competitors.

Skype and WhatsApp disrupted the telecom industry by using the Internet as a free infrastructure for calls and messages, while telecoms incurred heavy capital expenditures.

5. Others Who Do the Work.
How much does your business model get customers or third parties to create value for you for free?

All the value created in my business model is created for free by external parties.

10
○ ○ ○ ○ ○ ○ ○ ○ ○ ○ ○
0

I incur costs for all the value created in my business model.

Most of the value in Facebook's business model comes from content produced for free by more than 1 billion users. Similarly, merchants and shoppers create value for free for credit card companies.

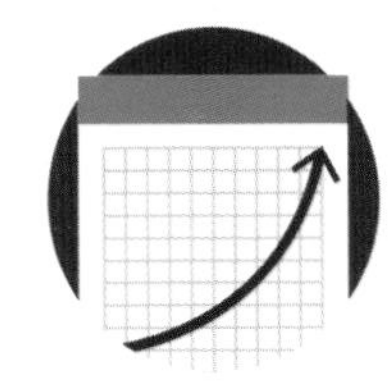

6. Scalability.
How easily can you grow without facing roadblocks (e.g., infrastructure, customer support, hiring)?

My business model has virtually no limits to growth.

10
○ ○ ○ ○ ○ ○ ○ ○ ○ ○ ○
0

Growing my business model requires substantial resources and effort.

Licensing and franchising are extremely scalable, as are platforms like Facebook or WhatsApp that serve hundreds of millions of users with few employees. Credit card companies are also an interesting example of scalability.

7. Protection from Competition.
How much is your business model protecting you from your competition?

My business model provides substantial moats that are hard to overcome.

10
○ ○ ○ ○ ○ ○ ○ ○ ○ ○ ○
0

My business model has no moats, and I'm vulnerable to competition.

Powerful business models are often hard to compete with. Ikea has found few imitators. Similarly, platform models like Apple with the App Store provide powerful moats.

2.6

Designing in Established Organizations

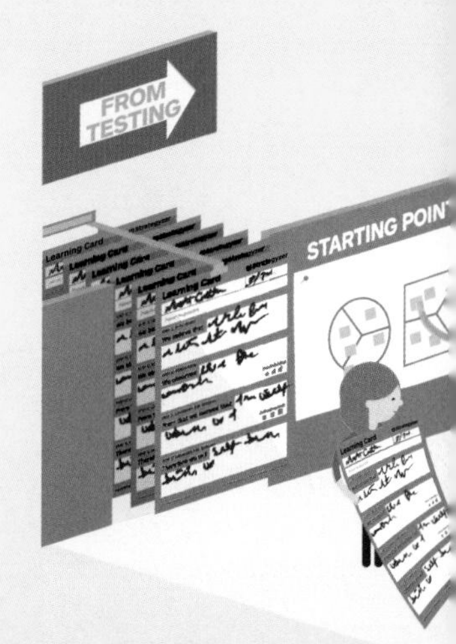

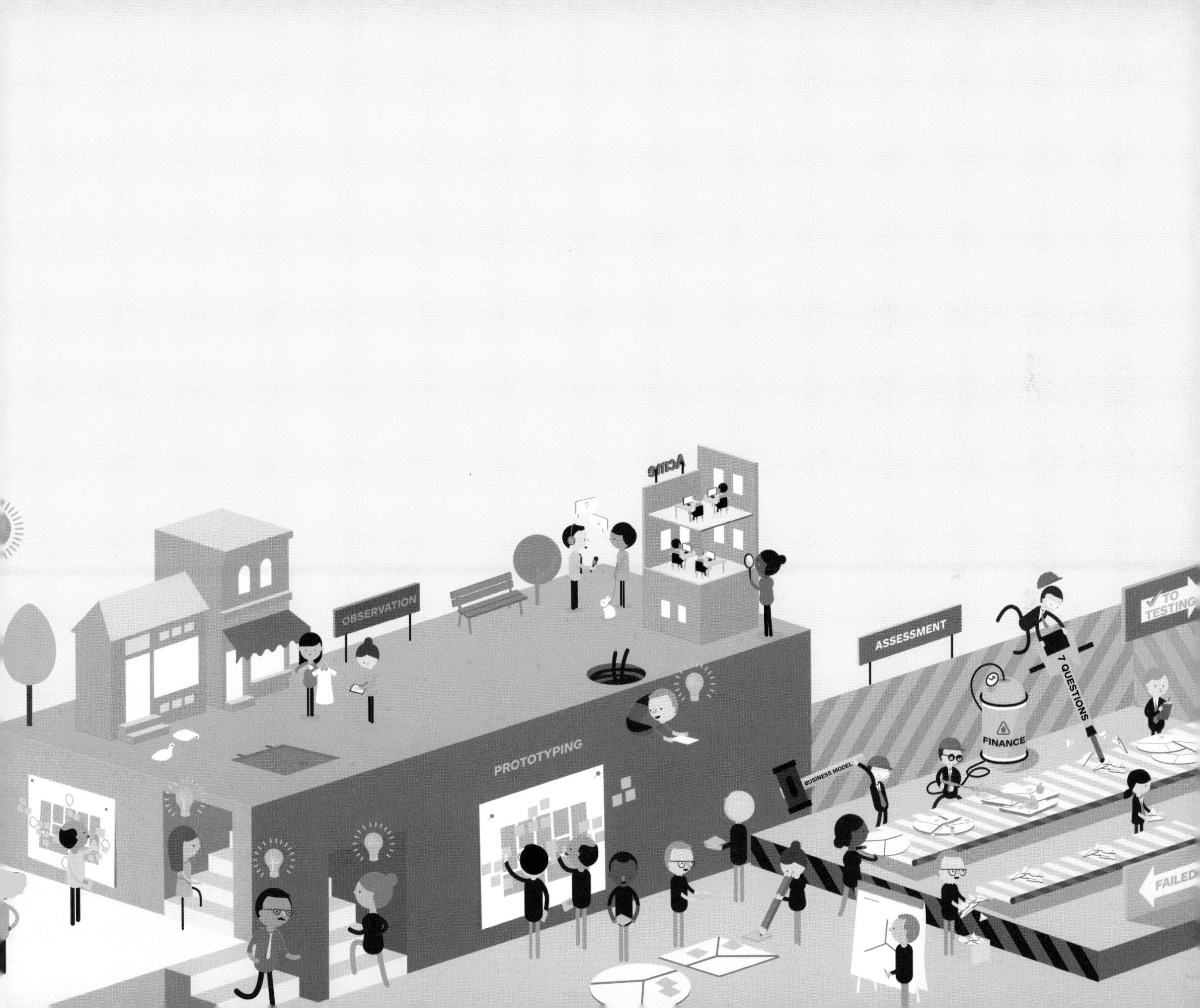
Acme
OBSERVATION
ASSESSMENT
TO TESTING
PROTOTYPING
7 QUESTIONS
FINANCE
BUSINESS MODEL
FAILED

Adopt the Right Attitude to Invent or Improve

Existing organizations need to improve existing value propositions and create new ones proactively. Make sure you understand on which end of the spectrum you are at the beginning of a particular project, because each requires a different attitude and process. Great companies will have a balanced portfolio of projects covering the entire spectrum from improve to invent.

invent

Objective	Design new value propositions regardless of the potential constraints given by existing value propositions and business models (although leadership may define other constraints).
Helps With	• Proactive bet on the future • Take on a crisis • Emergence of a game-changing technology, regulation, etc. • Response to a disruptive value proposition of a competitor
Financial Goals	At least 50 percent annual revenue growth (caveat: company-specific)
Risk and Uncertainty	High
Customer Knowledge	Low, potentially nonexistent
Business Model	Requires radical adaptions or changes
Attitude to Failure	Part of learning and iteration process
Mind-set	Open to exploring new possibilities
Design Approach	Radical/disruptive change to value proposition (and business model)
Main Activities	Search, test, and evaluate
Examples	*Amazon Web Services* Design of a new IT infrastructure value proposition targeted at a new customer segment. Builds on existing key resources and activities but requires a substantial expansion of Amazon.com's business model.

improve

Improve your existing value proposition(s) without radically changing or affecting the underlying business model(s).

- Renew outdated products and services.
- Ensure or maintain fit.
- Improve profit potential or cost structure.
- Keep growth going.
- Address customer complaints.

0 to15 percent annual revenue increase or more (caveat: company-specific)

Low

High

Little change

Not an option

Focused on making one or several aspects better

Incremental change and tweaks to existing value proposition

Refine, plan, and execute

Amazon Prime

Introduce a membership with special benefits targeted at frequent users of Amazon.com.

In between: Extend

A common situation in the Improve-Invent spectrum is the need to find new growth engines without investing in substantial changes to the existing business model. This is often required to monetize investments in existing models and platforms.

The objective is to search for new value propositions that substantially extend the existing underlying business model, without modifying too many aspects of it.

For example, with the introduction of the Kindle, Amazon created a new channel to extend its digital offering to Amazon.com customers. Although this presents a great new value proposition to its customers, it remains to a large extent within the parameters of its successfully established and well-mastered e-commerce business model.

Tip

Great companies manage a portfolio of value propositions and business models that cover the entire invent-improve spectrum and make synergies and competitive conflicts explicit. They are proactive and invent while they are still successful, rather than wait for a crisis.

The Business Book of the Future

Imagine if you were a business book publisher. How could you improve your present offering and invent the business book of the future, which might not even be a book anymore? We sketched out three ideas along the invent-improve spectrum.

The YouTube of business education
An online platform matching videos from business experts with customers who are looking for answers to their problems. This would require a substantial extension or reinvention of the business model of publishing books.

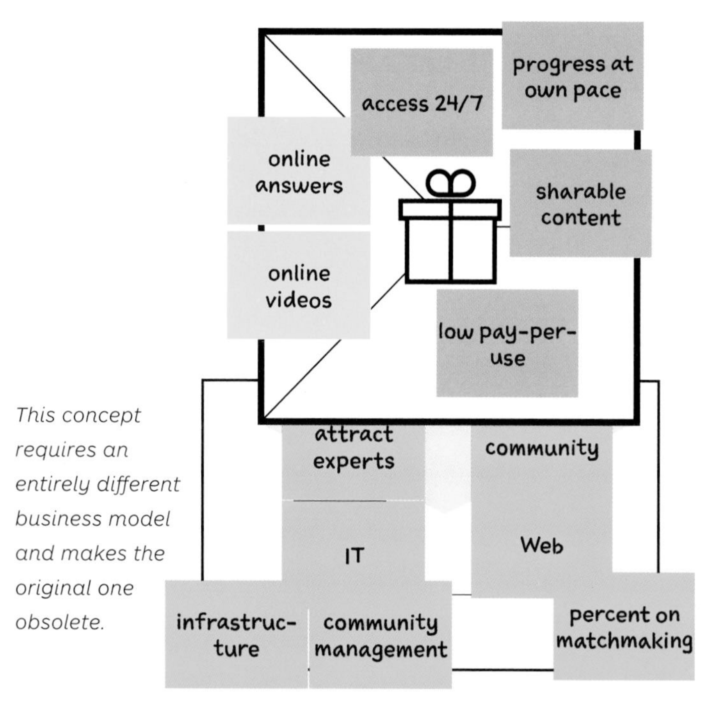

This concept requires an entirely different business model and makes the original one obsolete.

The 1-800-Business-Book hotline
A hotline number extending physical business books and offering on-demand answers. This would build on the existing business model but require an extension from a sales to a service model.

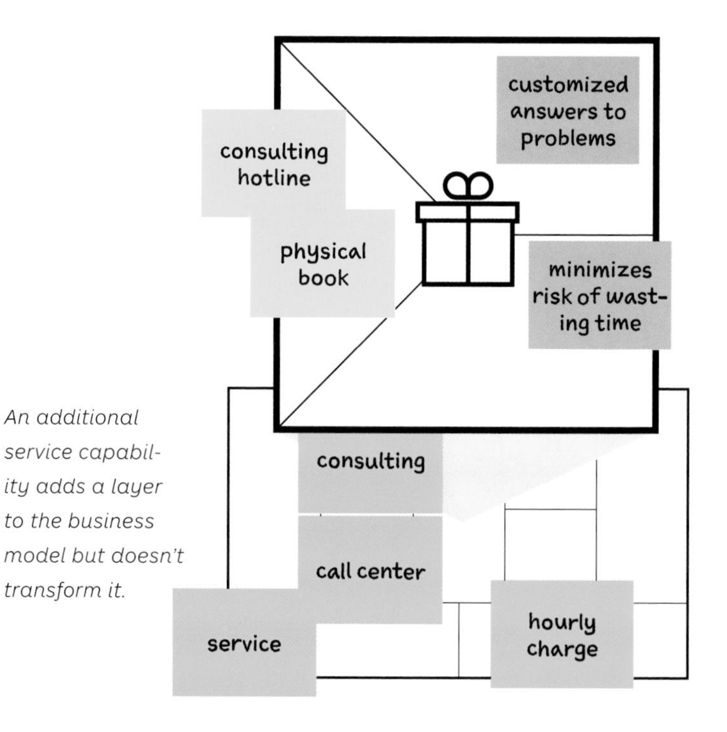

An additional service capability adds a layer to the business model but doesn't transform it.

improve

The practical business book
Improve business books by making them more visual and applicable without altering the core business model behind it substantially.

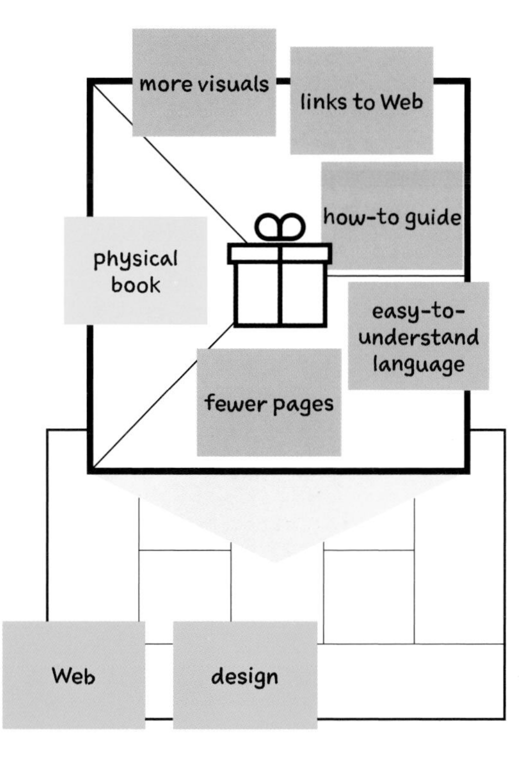

Improvements add to the value proposition and require only minor tweaks to the business model.

The more you move toward the invent end of the spectrum, the more your new value proposition will differ from your existing ones. Inventing new value propositions provides an opportunity to more closely address jobs that really matter to customers (in this case, getting answers to business questions).

Our three-tier value proposition consists of a physical book, sharable practical content online, and advanced learning through our online course. It is our attempt to push the boundaries of business learning and doing.

The value proposition of this book combined with online exercises and material on Strategyzer.com is our attempt to more closely address the jobs we believe matter to our readers.

Reinvent by Shifting from Products...

Construction equipment manufacturer Hilti reinvented its value proposition and business model by shifting from products to services. Its move from selling branded machine tools to guaranteeing timely access to them required a substantial overhaul not just of their value proposition but also of their business model. Let's learn how Hilti did it.

Many organizations aspire to regain a competitive advantage by transforming from a product manufacturer into a service provider. This requires a substantial reinvention.

An Expired Model

Hilti's old model focused mainly on selling high-quality machine tools directly to builders. They were known for breaking less often, lasting longer, and overall being less costly by minimizing time loss. Hilti tools also have a reputation for being particularly safe and enjoyable to work with.

Unfortunately, this old model was one of decreasing margins and subject to competition from lower-cost competitors.

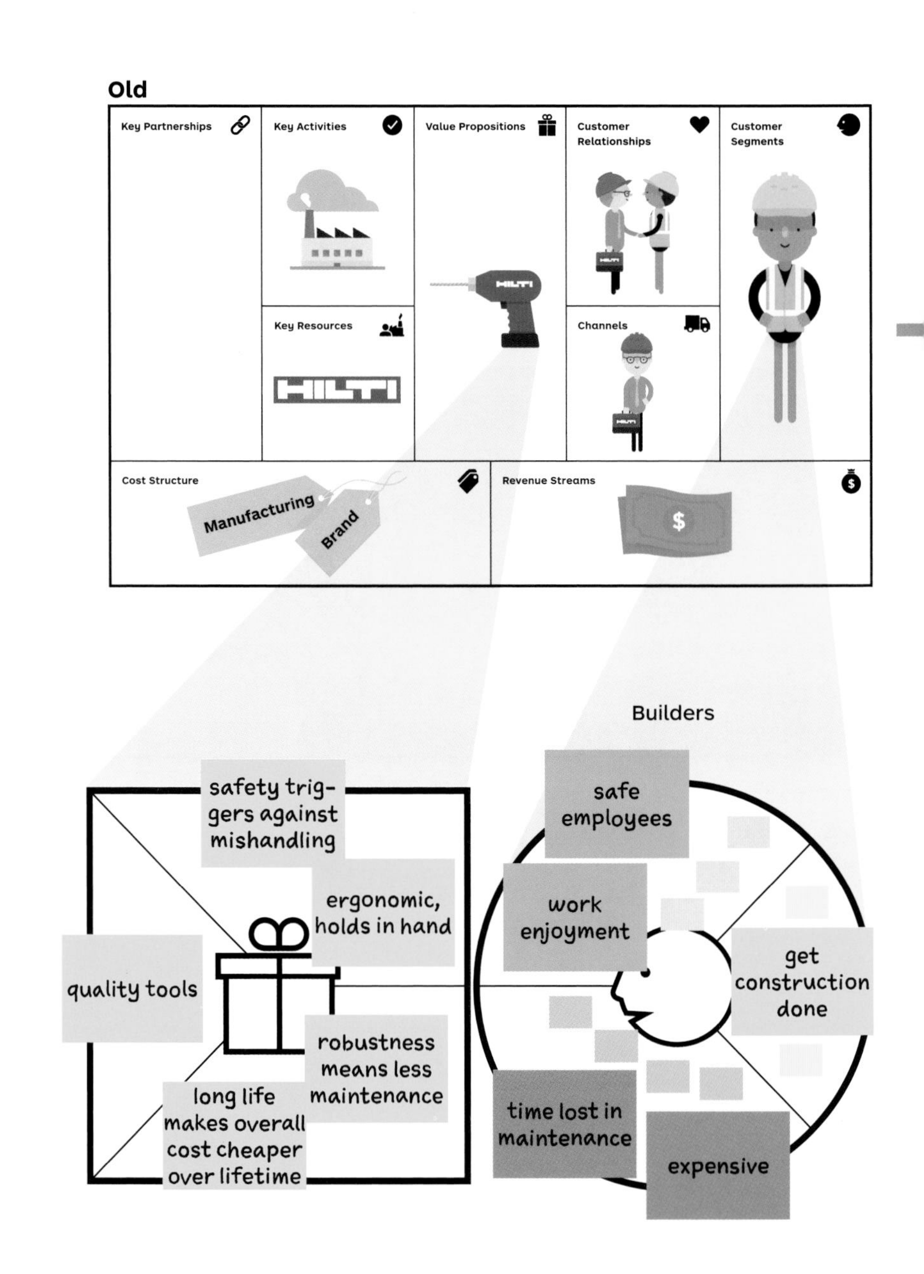

Read more about Hilti in Johnson, Seizing the Whitespace, 2010.

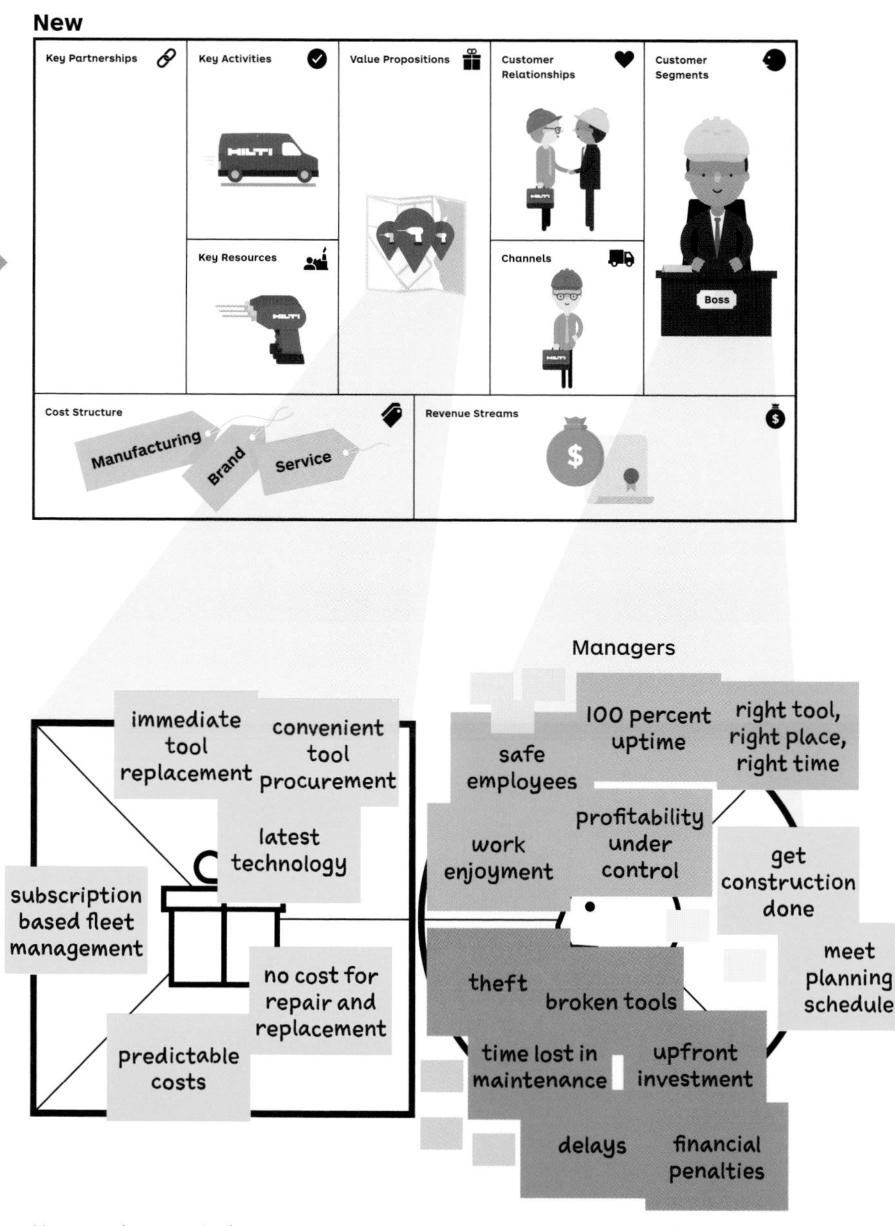

New service created: monthly subscriptions to fleet management utility

"New" customer, more important job identified: delivering on time!

...to Services

Hilti focused on a new job to be done after discovering that its tools were related to a more important customer job: that of delivering projects on time to avoid financial penalties. They learned that broken, malfunctioning, or stolen tools could lead to major delays and penalties. From there, Hilti moved toward a new value proposition, offering services around machine tools.

A Fresh Start

Hilti used its new service-based value proposition to create more value for construction companies by ensuring that they had the right tools at the right place at the right time. This would help construction companies achieve a much more predictable cost management and keep operations profitable.

Impact on the Business Model

Moving from products to services sounds like an easy and obvious value proposition shift, but it requires substantially reengineering the business model. Hilti had to add substantial new service resources and activities in addition to manufacturing. But it was worth it. With their new value proposition, Hilti achieves higher margins, recurring revenues, and better differentiation.

The Perfect Workshop Setting

Workshops are an important part of value proposition design in established organizations. Great workshops can make a big difference in the design process and lead to better results. The questions below will help you create the perfect setting.

Use sticky notes to move ideas around—ideally in several colors for color-coding.

Use thick markers so ideas are visible from afar.

Use wall-sized posters to sketch out big ideas.

Who should join?

Invite people with different backgrounds, especially when you know there will be a substantial effect on the business model. Their buy-in is crucial. Get customer-facing staff to participate to leverage their knowledge. Customers or partners may also be a good addition to help evaluate value propositions.

What should the format be?

As a rule of thumb, more viewpoints are generally better than fewer at the early stages of value proposition design. With 10 participants or more, you can explore several alternatives in parallel by working in groups of five. Smaller teams need to explore alternatives sequentially. At the later stages of developing and refining value propositions, fewer participants are usually better.

How can space be used as an instrument?

Great workshop spaces are an often-overlooked instrument to create outstanding workshops with exceptional outcomes. Choose a space that is sufficiently large and offers large walls or working areas. Set up the space to support creation, collaboration, and productivity. For breakthrough results, choose an unusual and inspiring venue.

What tools and materials are needed?

Prepare a self-service area with canvas posters, sticky notes, paper, blue tack, markers, and other tools so participants can help themselves with what they need.

Check readily available workshop material

Work-in-Progress Gallery/Inspiration Wall
Set up an area where you can expose canvases and other work in progress. Add an "inspiration wall" with content that participants can draw from, such as reference models, examples, and models of competitors.

Projector and Screen
This is used to show slides or customer videos. It should be easily viewable by all.

Small Group Areas
This is where work gets done. Four or five people per group is best. Do not use chairs or tables unless required for specific work. Keep working groups in the same room rather than a break-out room to retain high energy levels throughout the workshop.

Room Control
This space should be set aside for the facilitator and team to access computer, sound system, Wi-Fi, and maybe a printer.

Walls
Large vertical surfaces are indispensable, whether movable or part of the building. Make sure you can stick large posters, sticky notes, and flip chart paper on them.

Venue Size, Look, and Feel
As a rule of thumb, calculate $50m^2$ per 10 participants. Favor inspiring venues over boring hotel meeting rooms.

Plenary Space
Everyone can meet here for plenary presentations and discussions. It can be set up with or without tables.

Compose Your Workshop

A great workshop produces tangible and actionable outcomes. Use the tools and processes from this book to start designing a draft workshop outline that leads to great results.

Design Principles for a Great Workshop

- Create a workshop agenda with a clear thread that shows participants how the new or improved value proposition(s) or business model(s) will emerge.
- Take participants on a journey of many steps by focusing on one simple task (module) at a time.
- Avoid "blah blah blah" and favor structured interactions with tools like the canvases or processes like the thinking hats.
- Alternate between work in small groups (4–6 people) and plenary sessions for presentations and integration.
- Strictly manage time for each module, in particular for prototyping. Use a timer visible to all participants.
- Design the agenda as a series of iterations for the same value proposition (or business model). Design, critique, iterate, and pivot.
- Avoid slow activities after lunch.

Day 1

9 AM

10 AM

11 AM

12 PM

1 PM

2 PM

3 PM

4 PM

5 PM

Day 2

9 AM

10 AM

11 AM

12 PM

1 PM

2 PM

3 PM

4 PM

5 PM

Use the modules below as a menu of options to draft a workshop agenda.

Before Your Workshop
Do your homework and gather customer insights ➔ p. 106.

After Your Workshop
Get going with testing your value propositions and business models in the real world ➔ p. 172.

Get Online For:
- sample agendas
- templates and instructions
- all-in-one material package

Prototype Possibilities

Trigger Questions — ➔ p. 15, 17, 31, 33

CS Mapping — ➔ p. 22

VP Mapping — ➔ p. 36

Napkin Sketches — ➔ p. 80

Ad libs — ➔ p. 82

Flesh out Ideas with VPC — ➔ p. 84

Constraints — ➔ p. 90

New Ideas with Books — ➔ p. 92

Push / Pull Exercise — ➔ p. 94

Six Ways to Innovate — ➔ p. 102

Making Choices

Rank Jobs, Pains, and Gains — ➔ p. 20

Check Your Fit — ➔ p. 94

'Job' Selection — ➔ p. 100

10 Questions — ➔ p. 122

Voice of Customer — ➔ p. 124

Assess against Environment — ➔ p. 126

Differentiate from Competition — ➔ p. 128

De Bono's Hats — ➔ p. 136

Dotmocracy — ➔ p. 138

Selecting Prototype — ➔ p. 140

Back and Forth with Business Model

Back and Forth Iteration — ➔ p. 152

Numbers Projections — ➔ p. 154

7 BM Questions — ➔ p. 156

Preparing Tests

Extracting Hypotheses — ➔ p. 200

Prioritizing Hypotheses — ➔ p. 202

Test Design — ➔ p. 204

Choose a Mix of Experiments — ➔ p. 216

Test Road Map — ➔ p. 242–245

Breaks

Lunch

Coffee and snacks

Lessons Learned

Prototyping Possibilities

Rapidly prototype alternative value propositions and business models. Don't fall in love with your first ideas. Keep your early models rough enough to throw away without regret so that they can evolve and improve.

Understanding Customers

Imagine, observe, and understand your customers. Put yourself in their shoes. Learn what they are trying to get done in their work and in their lives. Understand what prevents them from getting this done well. Unearth which outcomes they are looking for.

Finding the Right Business Model

Search for the right value proposition embedded in the right business model, because every product, service, and technology can have many different models. Even the best value propositions can fail without a sound business model. The right business model can be the difference between success and failure.

The Value Proposition Designer
IAN

te

st

3

Reduce the risk and uncertainty of your ideas for new and improved value propositions by deciding What to Test p. 188. Then, get started with Testing Step-by-Step p. 196 and drawing from the Experiment Library p. 214 before Bringing It All Together p. 238 and measuring your progress.

HYPOTHESIS EXTRACTOR
Test Card
Strategyzer
HYPOTHESIS
#1
#2
#3
#4
#5
#6
#7
FROM DESIGN

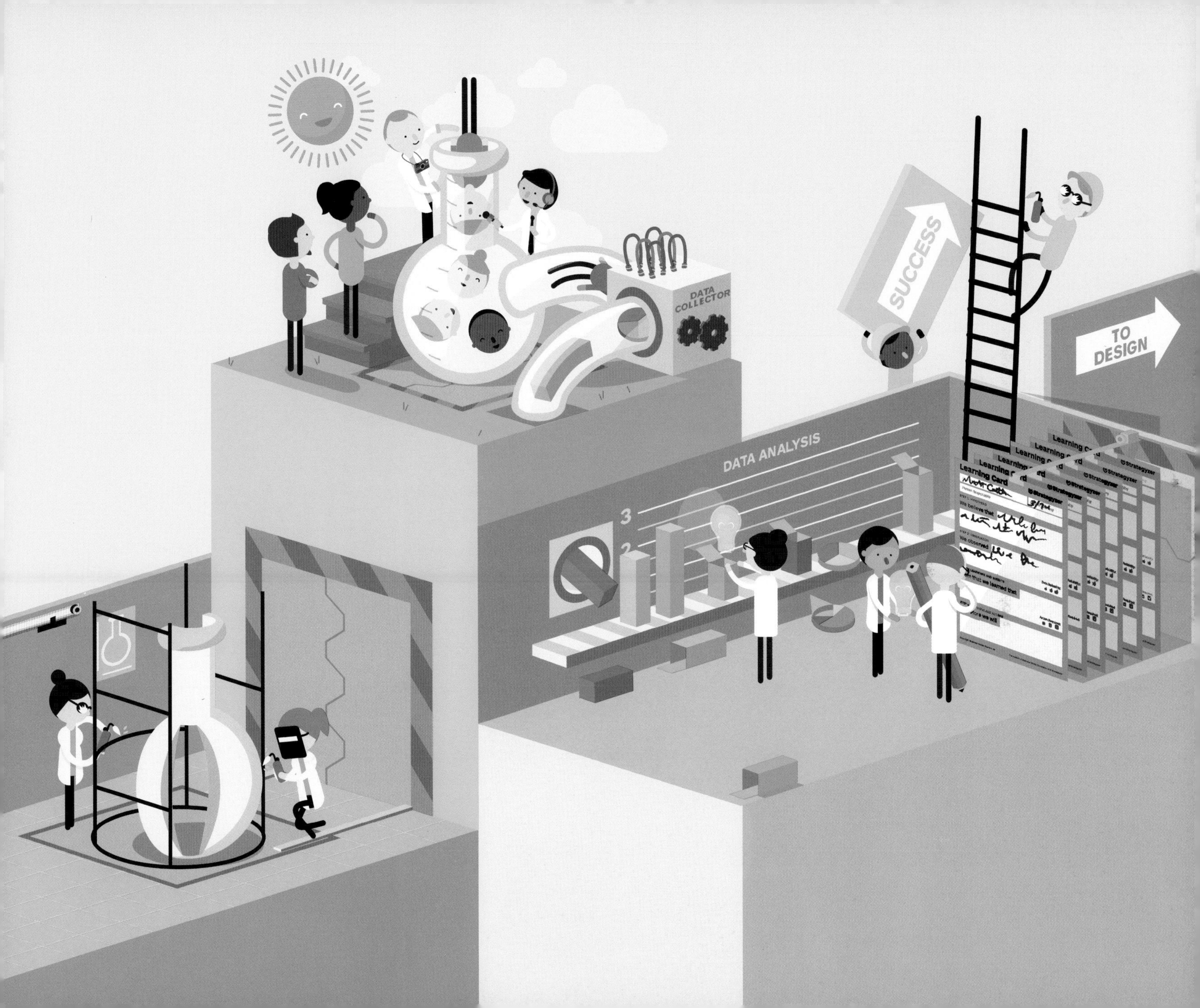

DATA COLLECTOR
SUCCESS
TO DESIGN
DATA ANALYSIS
3
2
Learning Card

Start Experimenting to Reduce Risk

When you start exploring new ideas, you are usually in a space of maximum uncertainty. You don't know if your ideas will work. Refining them in a business plan won't make them more likely to succeed. You are better off testing your ideas with cheap experiments to learn and systematically reduce uncertainty. Then increase spending on experiments, proto-types, and pilots with growing certainty. Test all aspects of your Value Proposition and Business Model Canvases, all the way from customers to partners (e.g., channel partners).

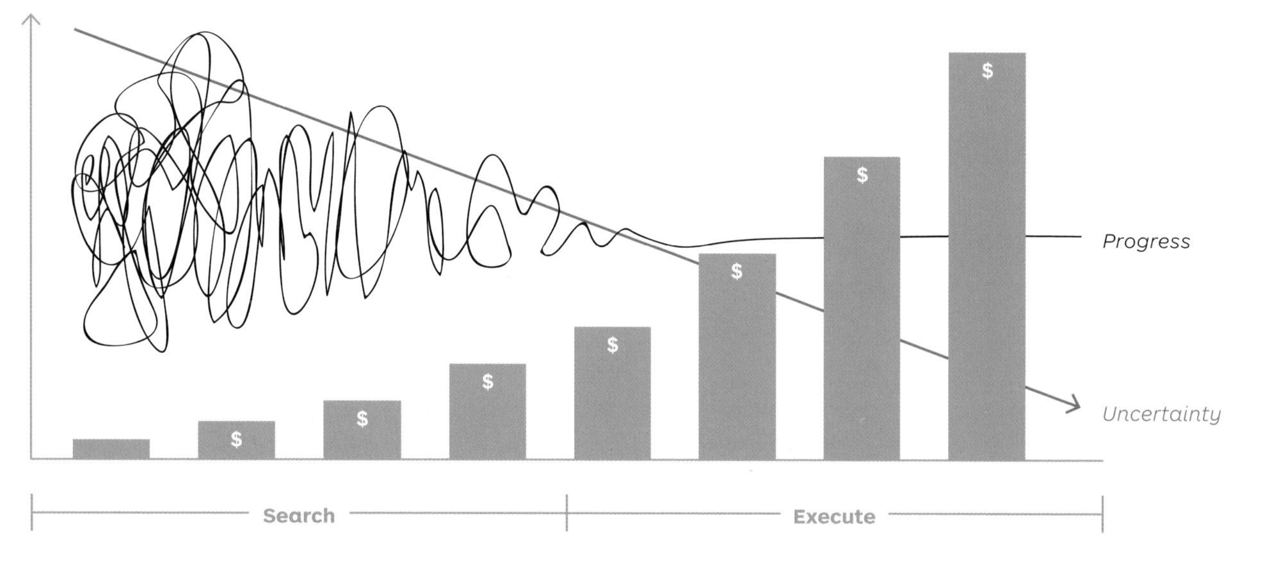

Business Plans vs. Experimentation Processes

The first step in any venture used to be writing a business plan. We now know better. Business plans are great execution documents in a known environment with sufficient certainty. Unfortunately, new ventures often take place under high uncertainty. Therefore, systematically testing ideas to learn what works and what doesn't is a far better approach than writing a plan. One might even argue that plans maximize risk. Their refined and polished nature gives the illusion that with great execution little can go wrong. Yet ideas dramatically change from inception to market readiness and often die along the way. You need to experiment, learn, and adapt to manage this change and progressively reduce risk and uncertainty. This process of experimentation, which we will explore on the following pages, is known as customer development and lean start-up.

Business Planning ⟷

	Applied to New Ventures	
We know	**Attitude**	Our customers and partners know
Business plan	**Tools**	Business Model and Value Proposition Canvas
Planning	**Process**	Customer development and lean start-up
Inside the building	**Where**	Outside the building
Execution of a plan	**Focus**	Experimentation and learning
Historical facts from past success	**Decision basis**	Facts and insights from experiments
Not addressed adequately	**Risk**	Minimized via learnings
Avoided	**Failure**	Embraced as means to learn and improve
Masked via detailed plan	**Uncertainty**	Acknowledged and reduced via experiments
Granular documents and spreadsheets	**Detail**	Dependent on level of evidence from experiments
Assumptions	**Numbers**	Evidence-based

10 Testing Principles

Apply these 10 principles when you start testing your value proposition ideas with a series of experiments. A good experimentation process produces evidence of what works and what doesn't. It also will enable you to adapt and change your value propositions and business models and systematically reduce risk and uncertainty.

Get "10 Testing Principles" poster

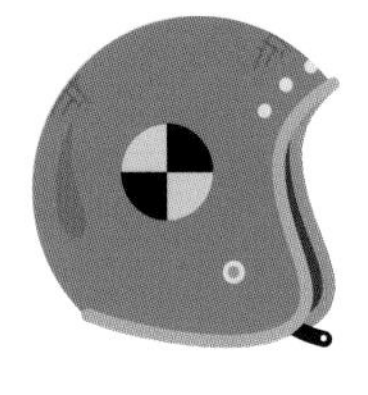

1

Realize that evidence trumps opinion. Whatever you, your boss, your investors, or anybody else thinks is trumped by (market) evidence.

2

Learn faster and reduce risk by embracing failure. Testing ideas comes with failure. Yet failing cheaply and quickly leads to more learning, which reduces risk.

3

Test early; refine later. Gather insights with early and cheap experiments before thinking through or describing your ideas in detail.

4

Experiments ≠ reality. Remember that experiments are a lens through which you try to understand reality. They are a great indicator, but they differ from reality.

5

Balance learnings and vision. Integrate test outcomes without turning your back on your vision.

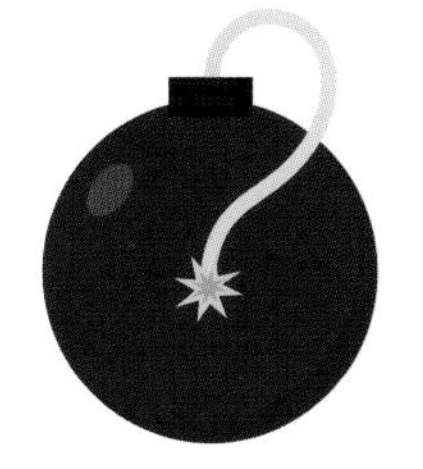

6

Identify idea killers.
Begin with testing the most important assumptions: those that could blow up your idea.

7

Understand customers first.
Test customer jobs, pains, and gains before testing what you could offer them.

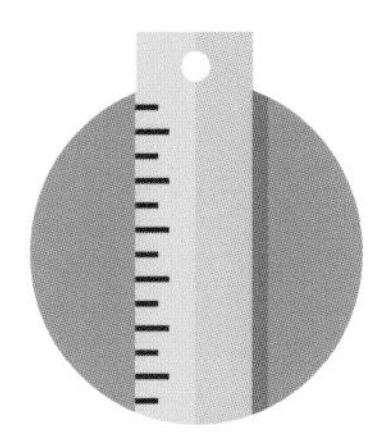

8

Make it measurable.
Good tests lead to measurable learning that gives you actionable insights.

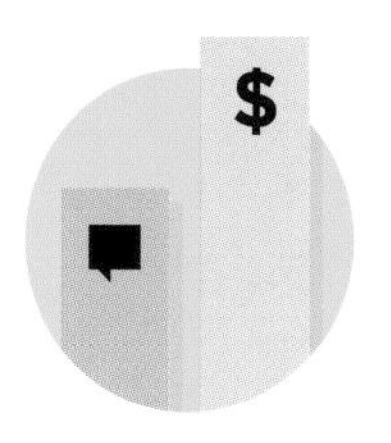

9

Accept that not all facts are equal.
Interviewees might tell you one thing and do another. Consider the reliability of your evidence.

10

Test irreversible decisions twice as much.
Make sure that decisions that have an irreversible impact are particularly well informed.

Introducing the Customer Development Process

Customer development is a four-step process invented by Steve Blank, serial entrepreneur turned author and educator. The basic premise is that there are no facts in the building, so you need to test your ideas with customers and stakeholders (e.g., channel partners or other key partners) before you implement them. In this book we use the customer development process to test the assumptions underlying Value Proposition and Business Model Canvases.

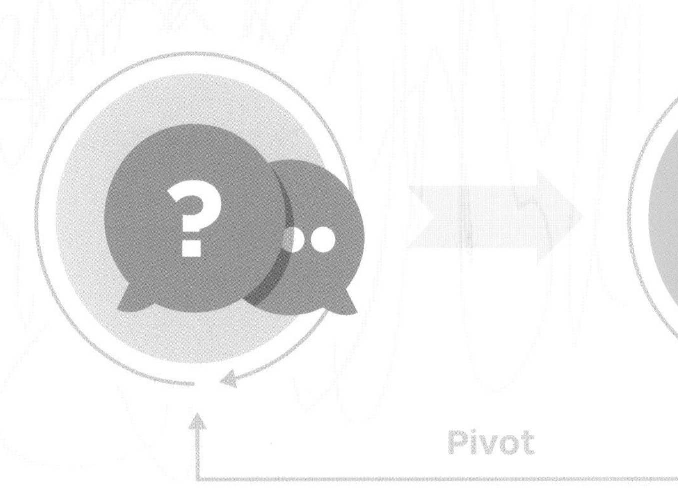

Customer Discovery
Get out of the building to learn about your customers' jobs, pains, and gains. Investigate what you could offer them to kill pains and create gains.

Customer Validation
Run experiments to test if customers value how your products and services intend to alleviate pains and create gains.

Search

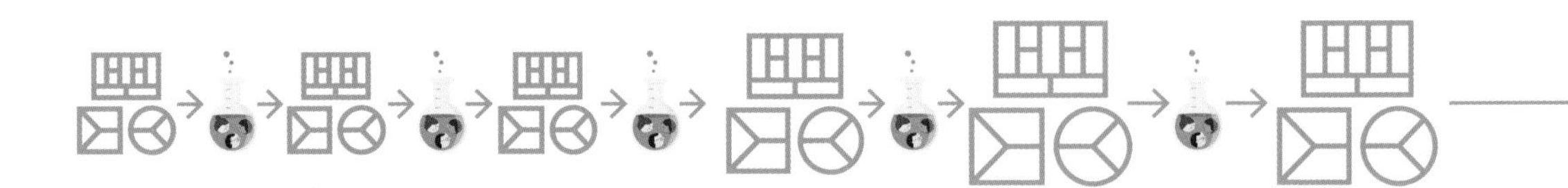

Customer Creation
Start building end user demand. Drive customers to your sales channels and begin scaling the business.

Company Building
Transition from a temporary organization designed to search and experiment to a structure focused on executing a validated model.

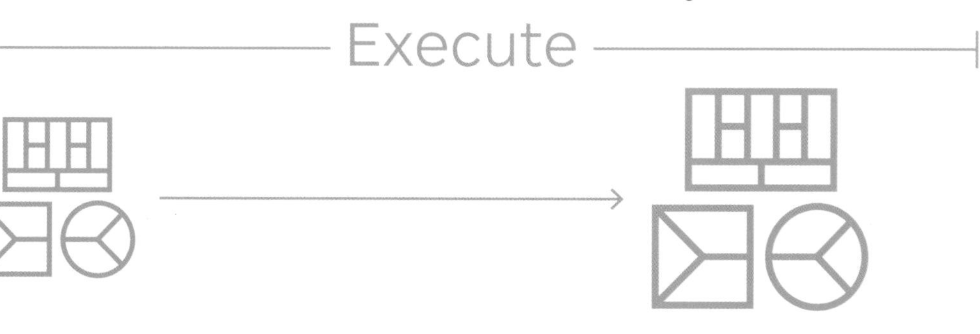

Search vs. Execute

The goal of the search phase is to experiment and learn which value propositions might sell and which business models could work. Your canvases will radically change and constantly evolve during this phase while you test every critical hypothesis. Only when you have validated your ideas do you get into the execution mode and scale. At the early stages of the process, your canvases change rapidly; they will stabilize with increasing knowledge from your experiments.

Tip

Capture every hypothesis, everything you tested, and everything you learned. Use the Value Proposition and Business Model Canvases to track your progress from initial idea and starting point toward a viable value proposition and business model. Keeping track of your progress and evidence produced along the way allows you to refer back to it if necessary.

Blank & Dorf ,The Startup Owner's Manual, 2012.

Integrating Lean Start-up Principles

Eric Ries launched the Lean Startup movement based on Steve Blank's customer development process. The idea is to eliminate slack and uncertainty from product development by continuously building, testing, and learning in an iterative process. Here we apply the three steps in combination with the canvases and customer development to test ideas, assumptions, and so-called minimum viable product (MVPs).

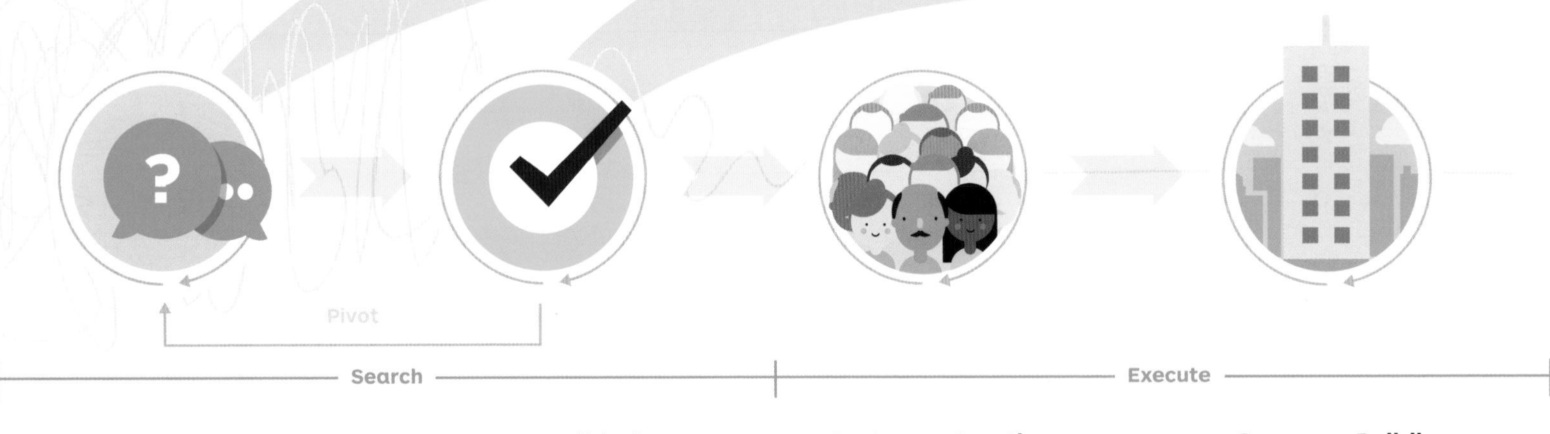

0. Generate a hypothesis.
Start with the Value Proposition and Business Model Canvases to define the critical hypotheses underlying your ideas in order to design the right experiments.

1. Design/build.
Design or build an artifact specifically conceived to test your hypotheses, gain insights, and learn. This could be a conceptual prototype, an experiment, or simply a basic prototype (MVP) of the products and services you intend to offer.

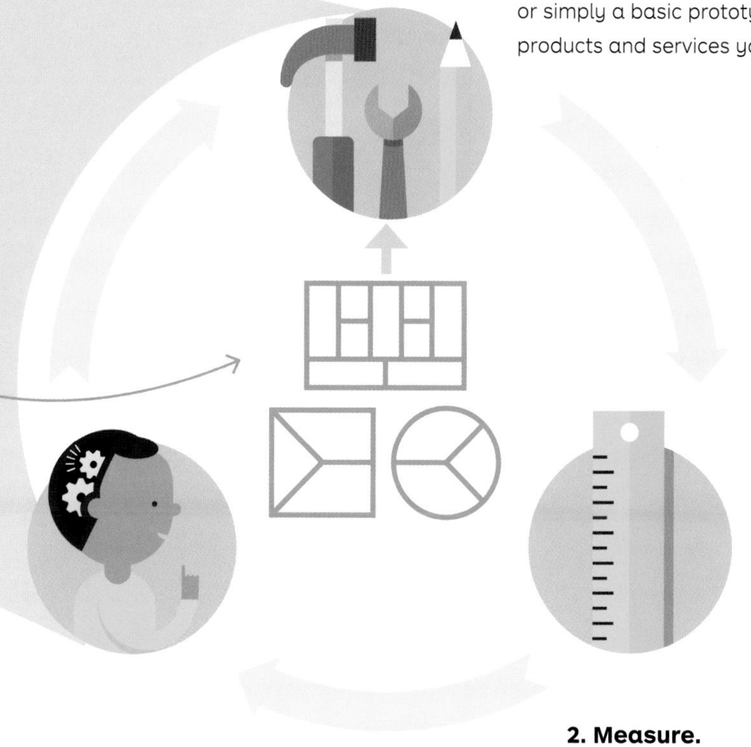

2. Measure.
Measure the performance of the artifact you designed or built.

3. Learn.
Analyze the performance of the artifact, compare to your initial hypotheses, and derive insights. Ask what you thought would happen. Describe what actually happened. Then outline what you will change and how you will do so.

Ries, The Lean Startup, 2011.

Apply Build, Measure, Learn

Apply the Lean Startup circle to more than just products and services. Use the same three steps of designing/building, testing/measuring, and learning with all the artifacts you create in *Value Proposition Design*. Apply design/build, test/measure, learn to your...

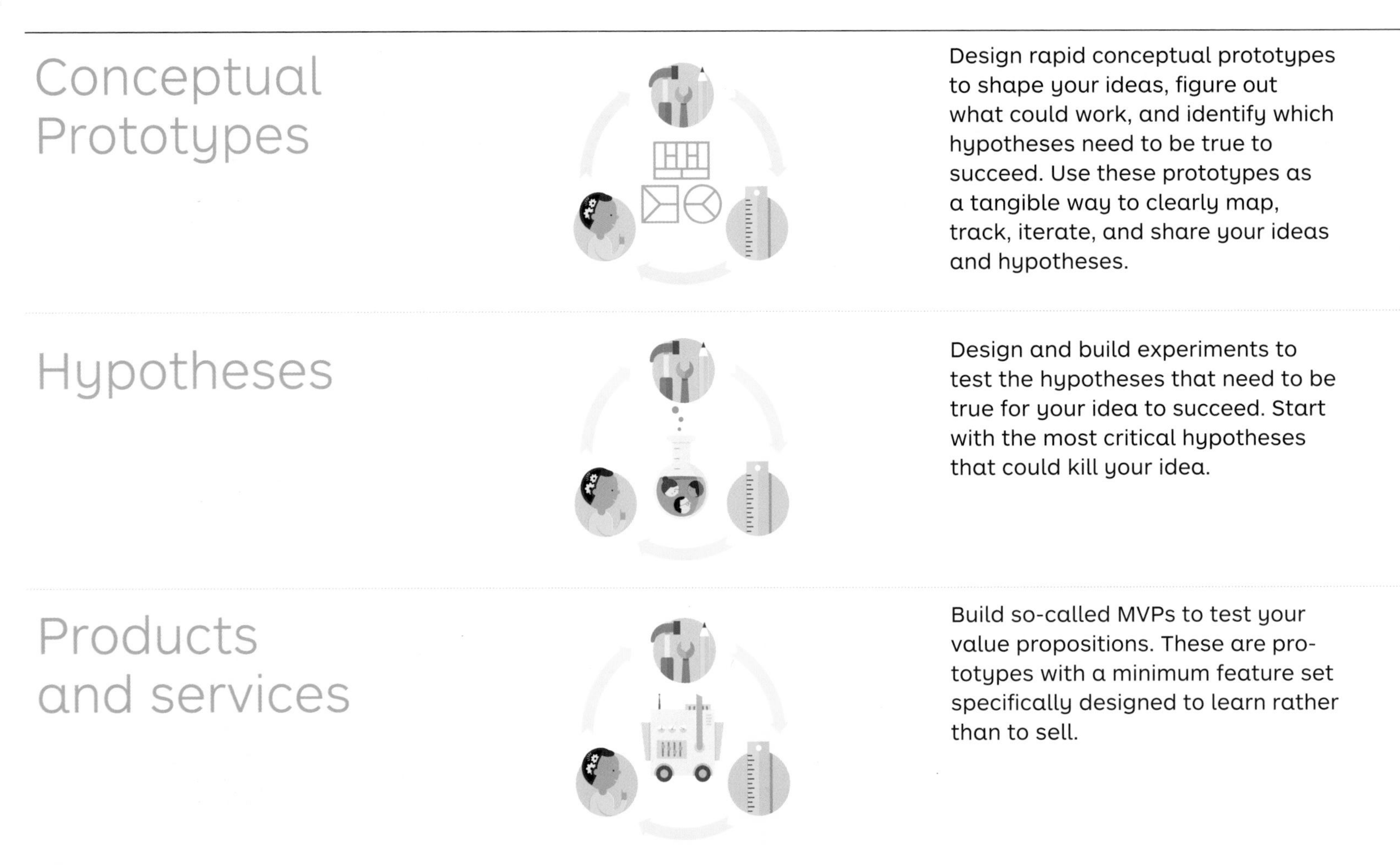

Conceptual Prototypes

Design rapid conceptual prototypes to shape your ideas, figure out what could work, and identify which hypotheses need to be true to succeed. Use these prototypes as a tangible way to clearly map, track, iterate, and share your ideas and hypotheses.

Hypotheses

Design and build experiments to test the hypotheses that need to be true for your idea to succeed. Start with the most critical hypotheses that could kill your idea.

Products and services

Build so-called MVPs to test your value propositions. These are prototypes with a minimum feature set specifically designed to learn rather than to sell.

Design/Build	Measure	Learn
Business Model and/ or Value Proposition Canvases to shape your ideas throughout the process	Performance of conceptual prototype: fit between customer profile and value map, ballpark figures, design assessment with 7 business model questions	If and why you need to adapt your conceptual prototypes Assumed financial performance of your business model Assumed fit Which hypotheses you need to test
Interviews, observations, and experiments to test initial value proposition and business model assumptions derived from conceptual prototyping	What actually happens in your experiments compared with what you thought would happen (i.e., your hypotheses)	If and why you need to change any of the building blocks of your Business Model or Value Proposition Canvas
MVPs with the benefits and features you want to test	If your products and service actually relieve pains and create gains for customers	If and why you need to change the products and services in your value proposition Which pain relievers and gain creators work and which ones don't

I present Shrek models, that's a Yiddish expression for making people nervous.
Frank Gehry, Architect

There are no facts in the building... So get the hell out and talk to customers.
Steve Blank, Entrepreneur & Educator

Fail early to succeed sooner.
David Kelley, Designer

3.1

What to Test

Testing the Circle

Prove which jobs, pains, and gains matter to customer most by conducting experiments that produce evidence beyond your initial customer research. Only after this has been done should you get started with your value proposition. This will prevent you from wasting time with products and services customers don't care about.

Provide evidence showing what customers care about (the circle) before focusing on how to help them (the square).

Start with jobs, pains, and gains

In the design section we looked at a series of techniques to better understand customers. In this chapter we go a step further. The objective of "testing the circle" is to confirm with evidence that our profile sketches, our initial research, our observations, and our insights from interviews were correct. We aim to know with more certainty which jobs, pains, and gains customers really care about.

Possessing evidence about customer jobs, pains, and gains before you focus on your value proposition is very powerful. If you start by testing your value proposition, you never know if customers are rejecting your value proposition or if you are simply addressing irrelevant jobs, pains, or gains. This is less likely to happen if you have evidence about which jobs, pains, and gains matter to customers.

Of course, this means you need to find creative ways to test customer preferences without already drawing on the use of minimum viable products (MVPs). We show how to do so with the tools in the testing library ➔ p. 214.

Provide evidence showing that your customers care about how your products and services kill pains and create gains.

The Art of Testing Value Propositions

It is an art to test how much your customers care about your value proposition because the goal is to do so as cheaply and quickly as possible without implementing the value proposition in its entirety.

You need to test your customers' taste for your products and services one pain reliever and gain creator at a time by designing experiments that are measurable, provide insights, and allow you to learn and improve ➔ p. 214.

Make sure your experiments allow you to understand which aspects of your products and services customers appreciate, so that you can avoid offering anything unnecessary. In other words, remove any features or efforts that don't contribute directly to the learning you seek.

Always make sure you aim to find the simplest, quickest, and cheapest way to test a pain reliever or gain creator before you start prototyping products and services.

Testing the Rectangle

Test the most critical assumptions underlying the business model your value proposition is embedded in. Remember, even great value propositions can fail without a sound business model. Provide evidence showing that your business model is likely to work, will generate more revenue than costs, and will create value not only for your customers but for your business.

Provide evidence showing that the way you intend to create, deliver, and capture value is likely to work.

Don't neglect testing your business model
You can fail even with a successful value proposition if your business model generates less revenue than it incurs costs. Many creators are so focused on designing and testing products and services that they sometimes neglect this obvious equation (profit = revenues - costs) resulting from the building blocks of the Business Model Canvas.

A value proposition that customers want is worth little if you don't have the channels to reach customers in a way they want to be reached. Likewise, a business model that spends more money on acquiring customers than it will earn from revenue from those same customers won't survive over the long term. Similarly, a company will obviously go out of business if resources and activities required to create value are more costly than the value they capture. In some markets you might need access to key partners who might not be interested in working with you.

Design experiments that address the most important things that have to be true for your business model to work. Testing such critical assumptions will prevent you from failing with a great value proposition that customers actually want.

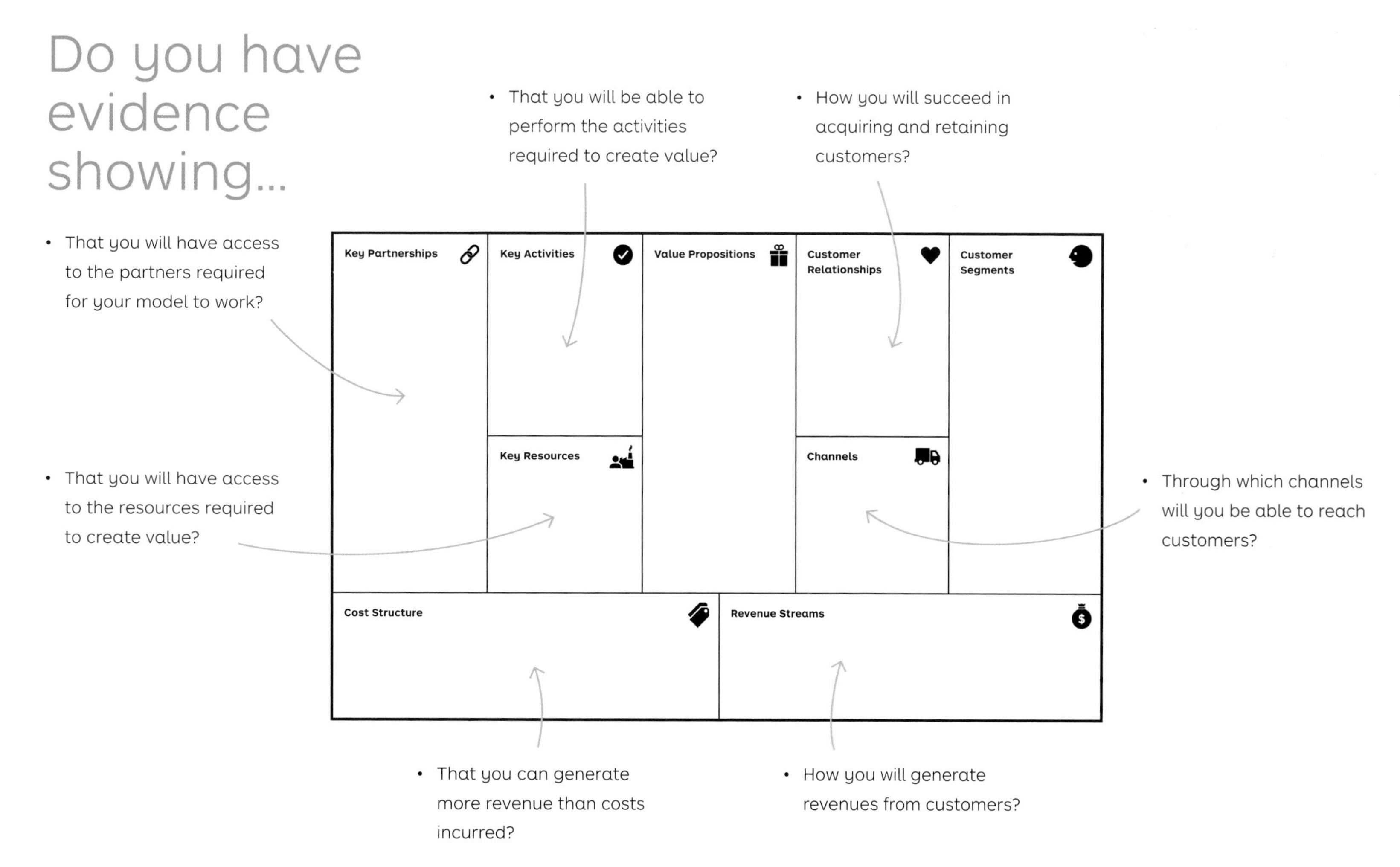
Do you have evidence showing…
That you will have access to the partners required for your model to work?
That you will be able to perform the activities required to create value?
How you will succeed in acquiring and retaining customers?
Key Partnerships
Key Activities
Value Propositions
Customer Relationships
Customer Segments
Key Resources
Channels
That you will have access to the resources required to create value?
Through which channels will you be able to reach customers?
Cost Structure
Revenue Streams
That you can generate more revenue than costs incurred?
How you will generate revenues from customers?

3.2

Testing Step-by-Step

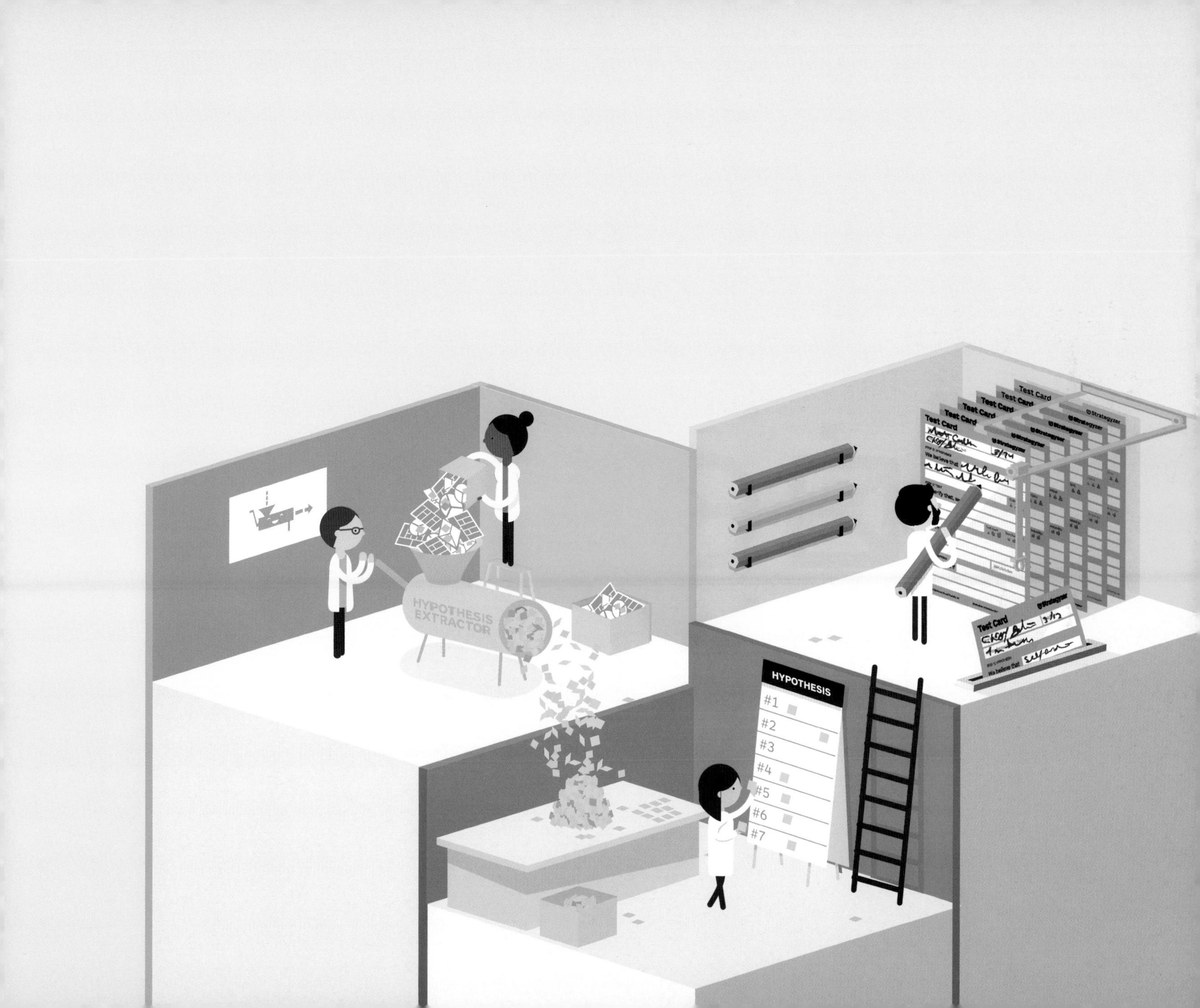
HYPOTHESIS EXTRACTOR
Test Card
Strategyzer
HYPOTHESIS
#1
#2
#3
#4
#5
#6
#7

Overview of the Testing Process

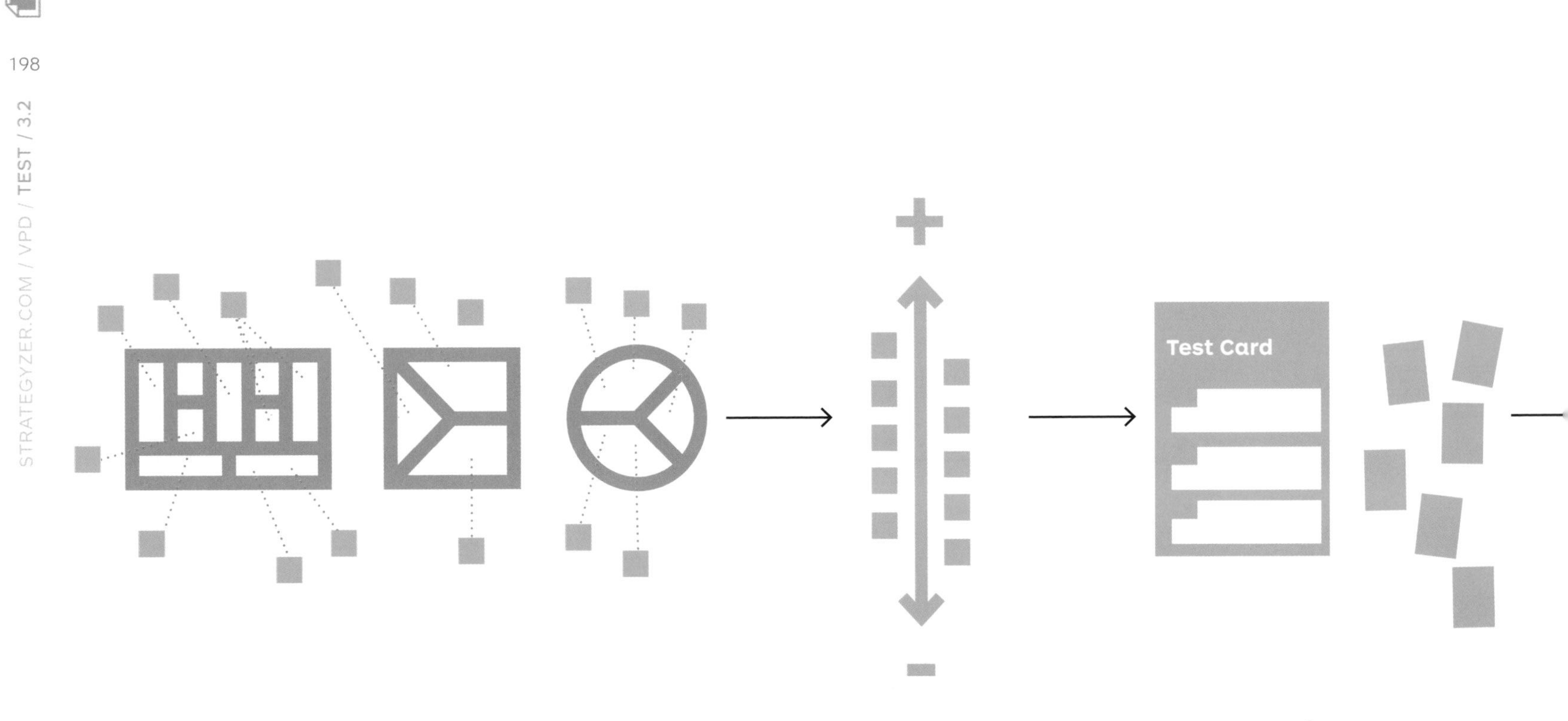

Extract Hypotheses

p. 200

Prioritize Hypotheses

p. 202

Design Tests

p. 204

Design Testing

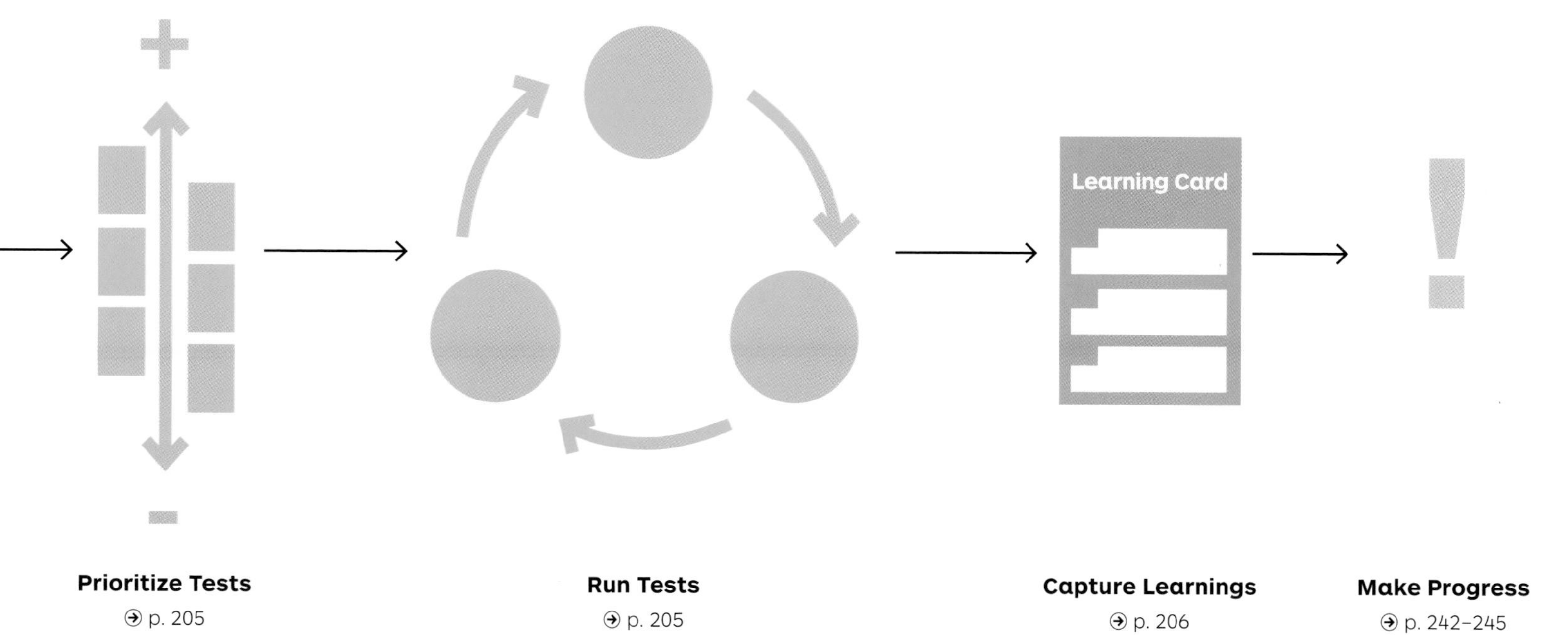

Prioritize Tests
p. 205

Run Tests
p. 205

Capture Learnings
p. 206

Make Progress
p. 242–245

Get "Testing Process Overview" poster

Extract Your Hypotheses: What Needs to Be True for Your Idea to Work?

Use the Value Proposition and Business Model Canvases to identify what to test before you "get out of the building." Define the most important things that must be true for your idea to work.

Do the exercise online

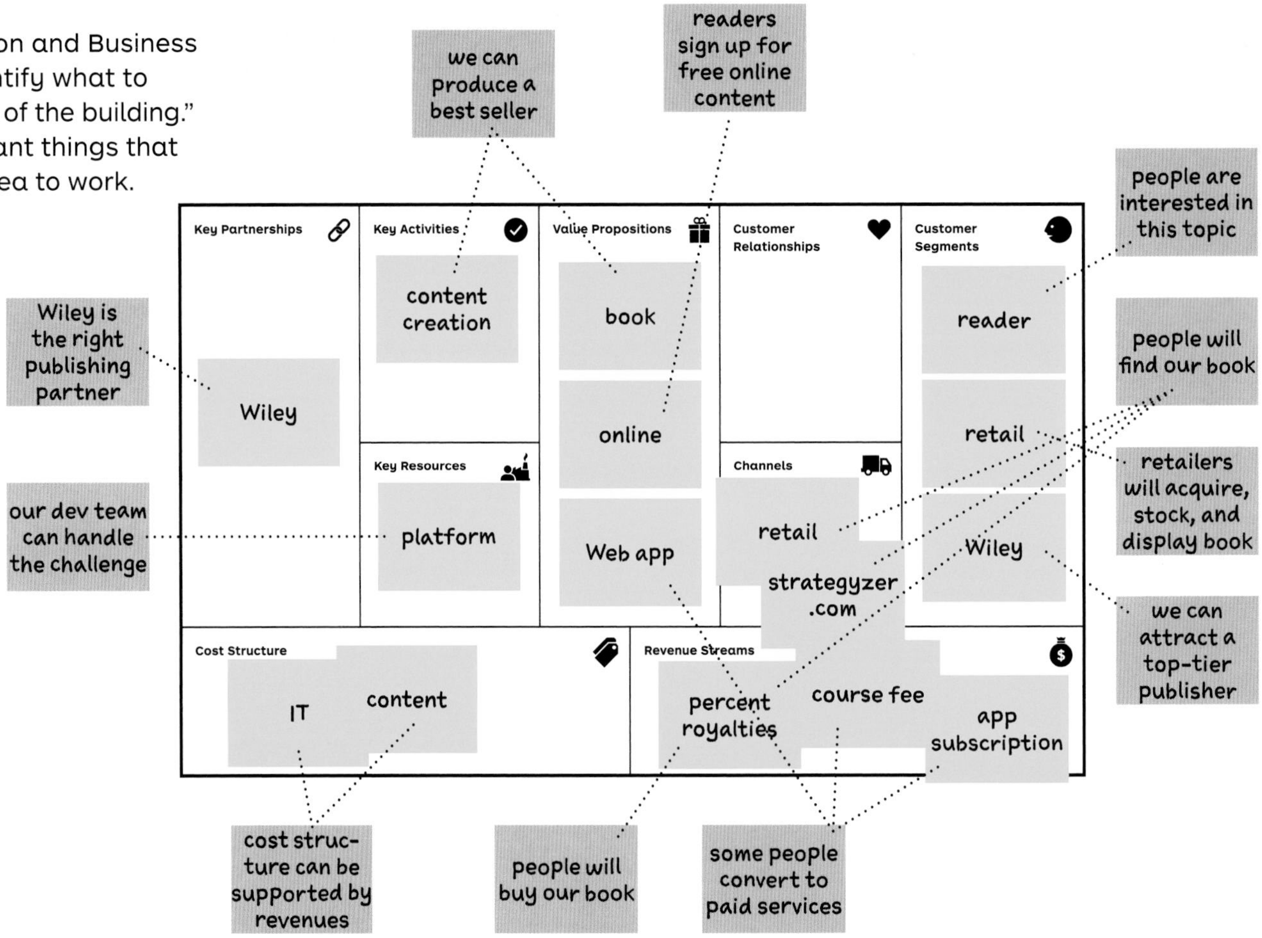

Key Partnerships
Wiley
Key Activities
content creation
Key Resources
platform
Value Propositions
book
online
Web app
Customer Relationships
Channels
retail
strategyzer .com
Customer Segments
reader
retail
Wiley
Cost Structure
IT
content
Revenue Streams
percent royalties
course fee
app subscription

we can produce a best seller
readers sign up for free online content
people are interested in this topic
Wiley is the right publishing partner
people will find our book
retailers will acquire, stock, and display book
our dev team can handle the challenge
we can attract a top-tier publisher
cost structure can be supported by revenues
people will buy our book
some people convert to paid services

To succeed, ask yourself what needs to be true about... → **...your business model?**

DEF·I·NI·TION

Business Hypothesis

Something that needs to be true for your idea to work partially or fully but that hasn't been validated yet.

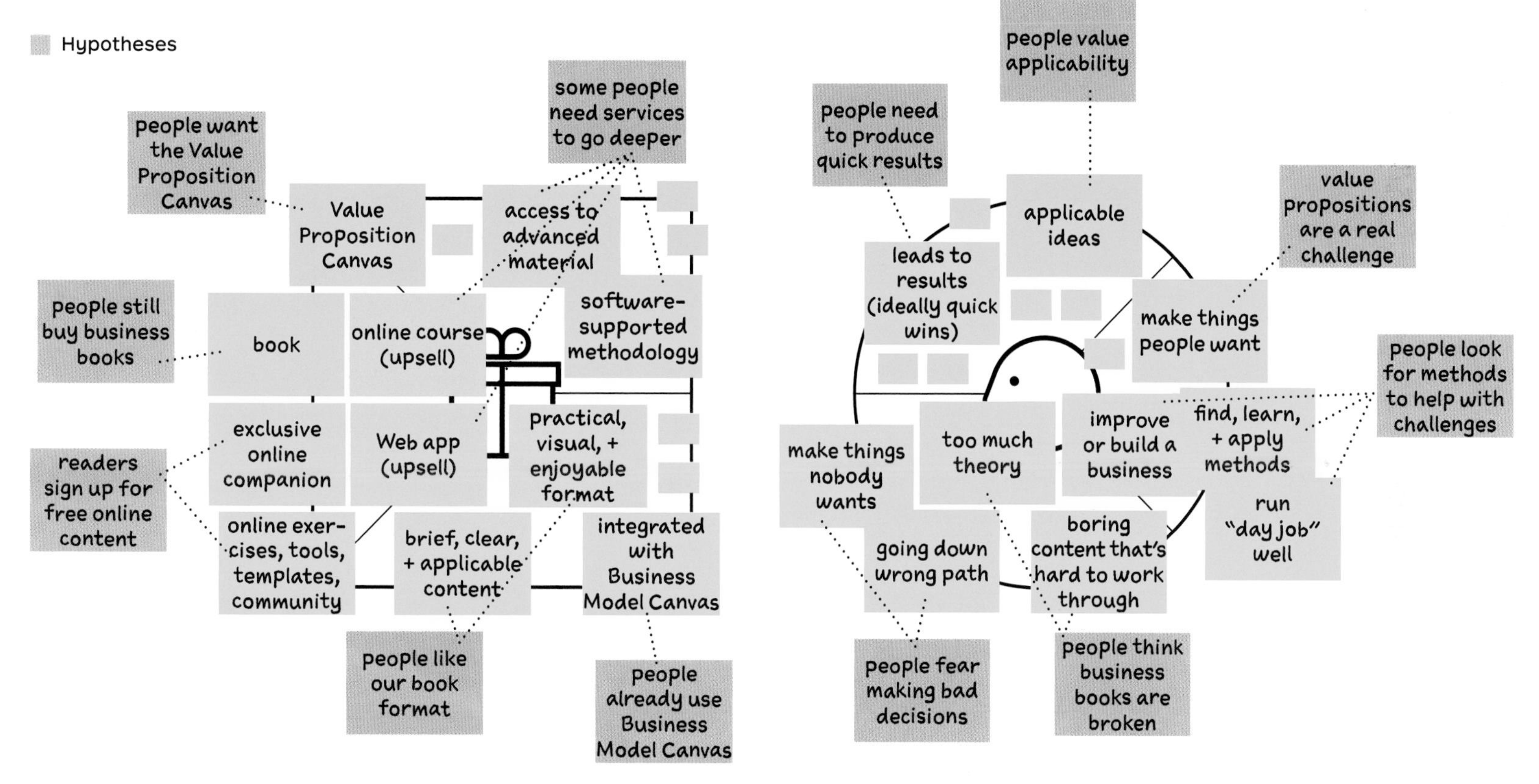

...your value proposition? **...your customer?**

Prioritize Your Hypotheses: What Could Kill Your Business

Not all hypotheses are equally critical. Some can kill your business, whereas others matter only once you get the most important hypotheses right. Start prioritizing what's critical to survival.

Identify the business killers. These are the hypotheses that are critical to the survival of your idea. Test them first!

Rank all your hypotheses in order of how critical they are for your idea to survive and thrive:

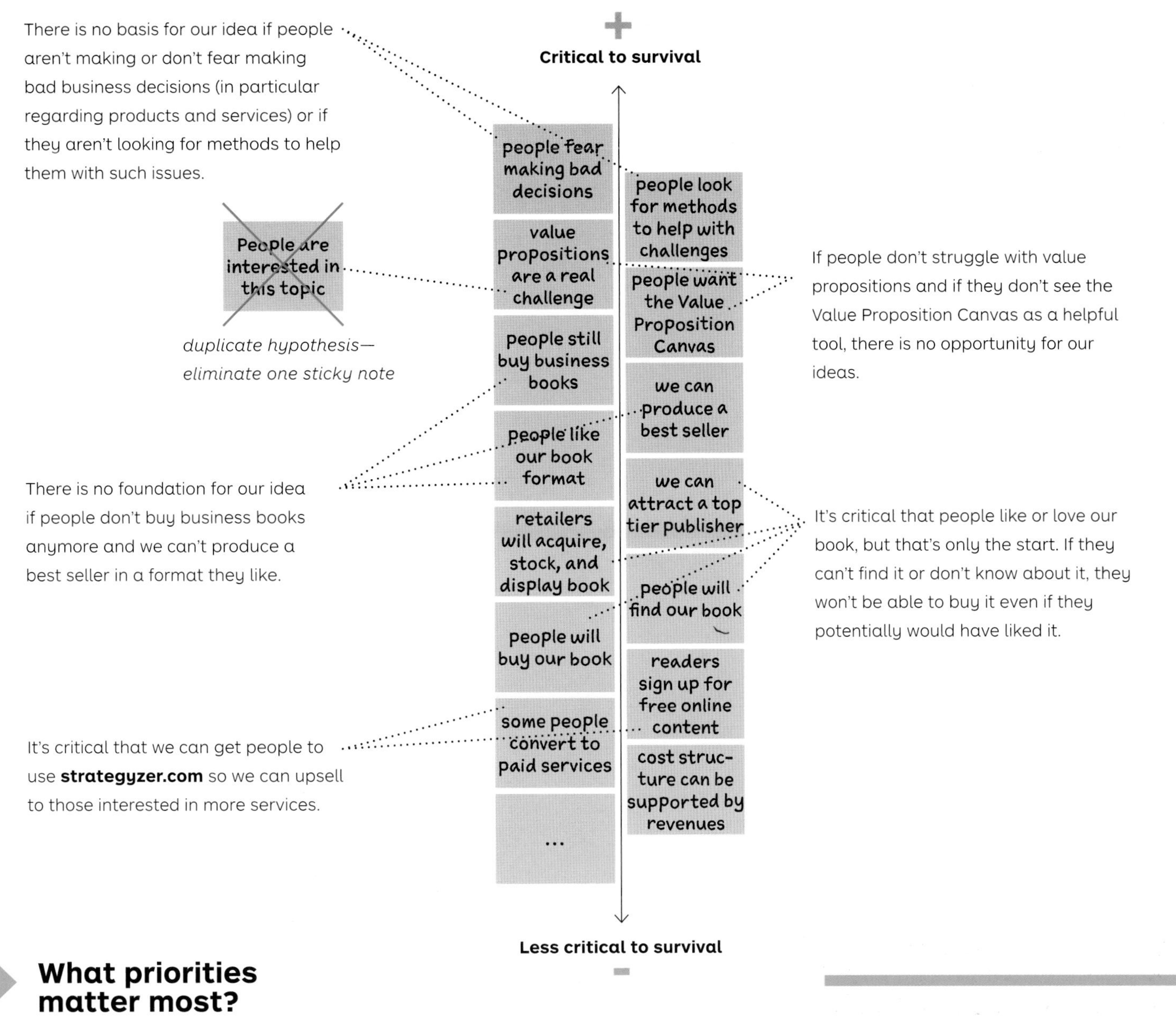

What priorities matter most?

Design Your Experiments with the Test Card

Structure all of your experiments with this simple Test Card. Start by testing the most critical hypotheses.

Download the Test Card and do the exercise online

1

Design an experiment.
Describe the hypothesis that you want to test.

Outline the experiment you are going to design to verify if the hypothesis is correct or needs to be rejected and revised.

Define what data you are going to measure.

Define a target threshold to validate or invalidate the tested hypothesis.
Caveat: Consider following up with additional experiments to increase certainty.

Test Card Strategyzer

AdWords campaign | May 1, 2014

Natasha Hanshaw | 2 weeks

STEP 1: HYPOTHESIS
We believe that businesspeople are looking for methods to help them design better value propositions.
Critical:

STEP 2: TEST
To verify that, we will launch a Google AdWords campaign around the search term "value proposition."
Test Cost: Data Reliability:

STEP 3: METRIC
And measure how the advertising campaign performs in terms of clicks.
Time Required:

STEP 4: CRITERIA
We are right if we can achieve a click-through rate (CTR) of at least 2 percent (number of clicks divided by total impressions of ad).

Copyright Business Model Foundry AG *The makers of Business Model Generation and Strategyzer*

Name the test, set a due date, and list the person responsible.

Indicate how critical this hypothesis is for the entire idea to work.

Indicate how costly this test will be to execute.

Indicate how reliable the measured data are.

Indicate how long it takes until this test produces results.

How will I learn?

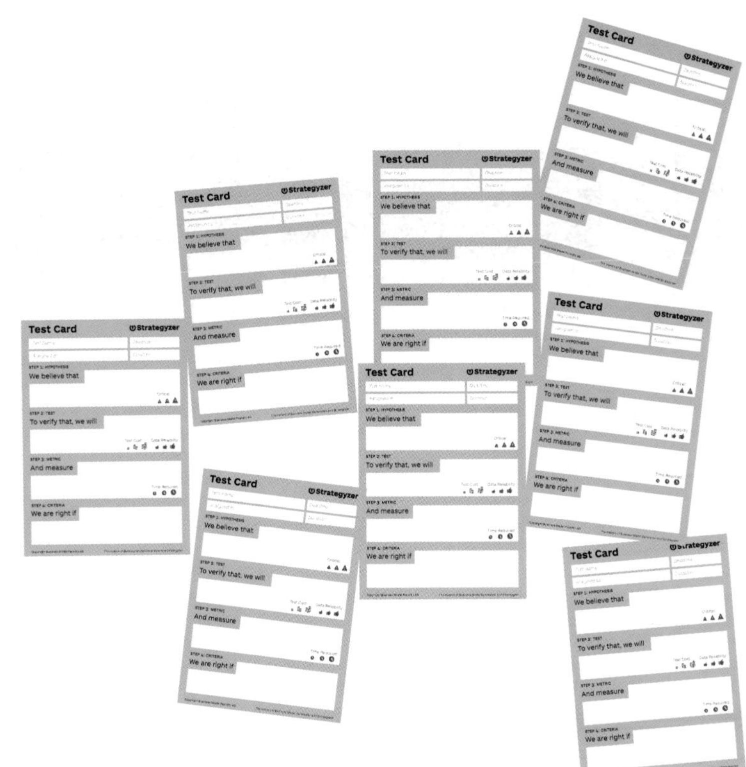

2

Design a series of experiments for the most critical hypotheses.

Tip

Consider testing the most critical hypotheses with several experiments. Start with cheap and quick tests. Then follow up with more elaborate and reliable tests if necessary. Thus, you may create several Test Cards for the same hypotheses.

3

Rank Test Cards.

Prioritize your Test Cards. Rank the most critical hypotheses highest, but prioritize cheap and quick tests to be done early in the process, when uncertainty is at its maximum. Increase your spending on experiments that produce more reliable evidence and insights with growing certainty.

4

Run experiments.

Start performing the experiments at the top of your list.

Caveat: If your first experiments invalidate your initial hypotheses, you might have to go back to the drawing board and rethink your ideas. This might render the remaining Test Cards in your list irrelevant.

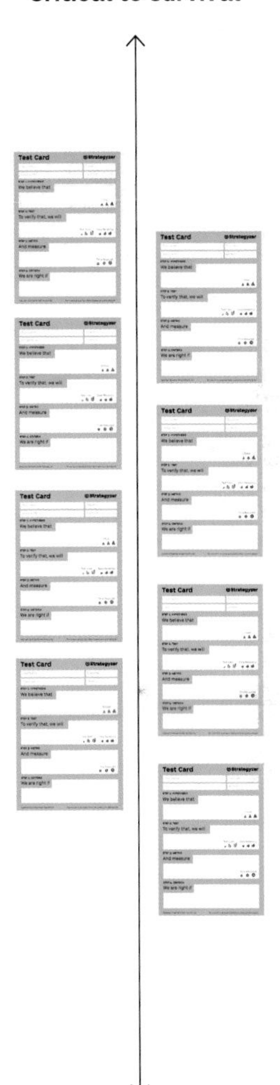

Repeat.

Where can I get the most learning the fastest?

Capture Your Insights with the Learning Card

Structure all of your insights with this simple Learning Card.

Describe the hypothesis that you tested.

Outline the outcomes of your experiment(s) in terms of data and results. A Learning Card may aggregate the observations from several Test Cards.

Explain what conclusions and insights you derived from the test results.

Describe what actions you will take based on your insights.

Download the Learning Card

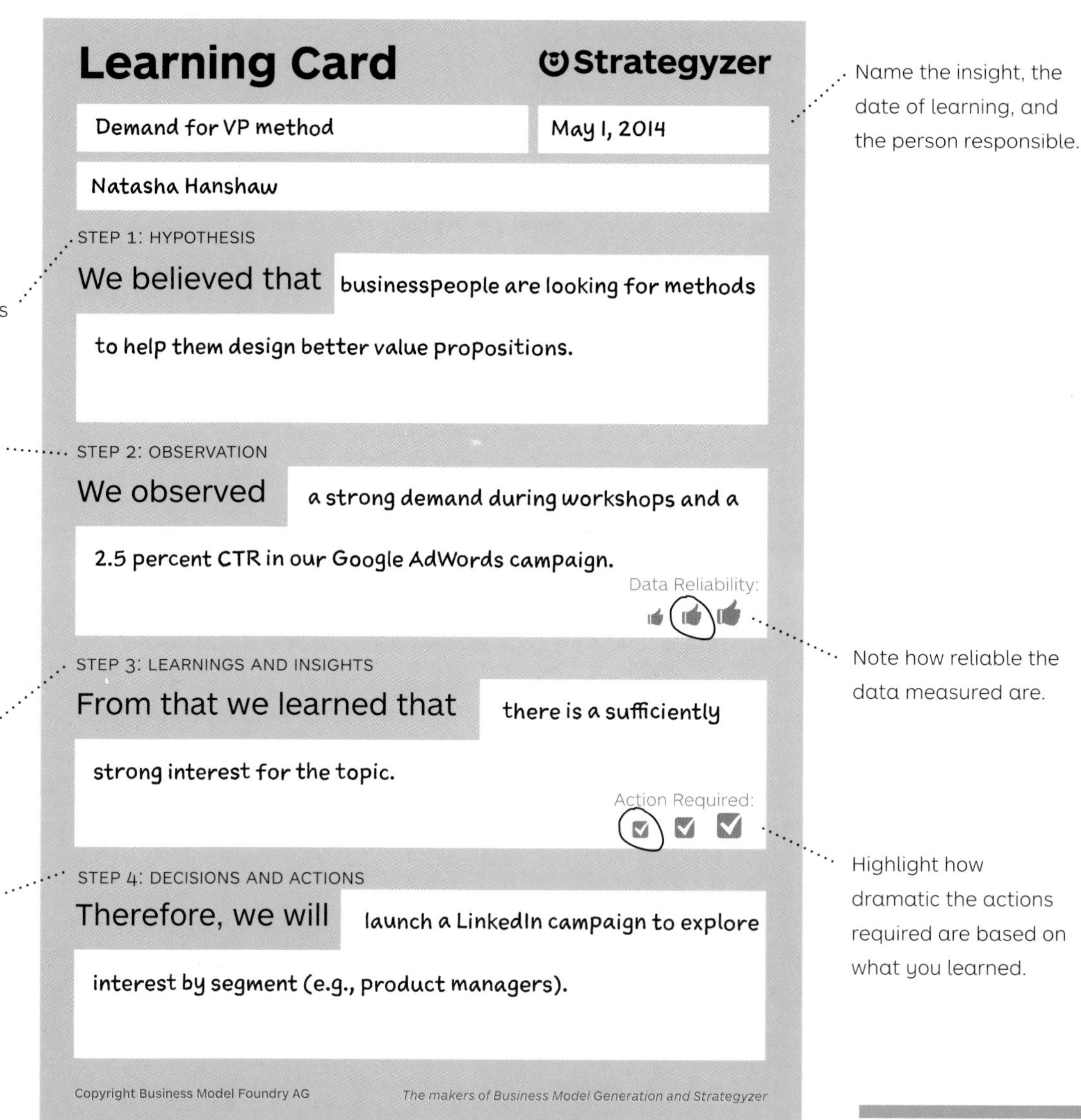

Name the insight, the date of learning, and the person responsible.

Note how reliable the data measured are.

Highlight how dramatic the actions required are based on what you learned.

Invalidated

Get back to the drawing board: pivot.
Find new alternative segments, value propositions, or business models to make your ideas work when your tests invalidate your first attempts.

For example, when you have invalidated customer interest for your value proposition around a novel technology, search for new potential customers, value propositions, and business models.

Learn more

Seek confirmation.
Design and conduct further tests when quick and early experiments based on a small amount of data indicate the need for drastic actions.

For example, if interviews with potential customers show a strong interest for a service that requires heavy investments to launch, follow up with research and experiments that produce more reliable data validating customer interest.

Deepen your understanding.
Design and conduct further tests to understand why a trend is taking place once you discovered that it is taking place.

For example, if the quantitative data of an experiment show that potential customers are not interested, follow up with qualitative interviews to understand why they are uninterested.

Validated

Expand to next building block.
Move on to test your next important hypothesis when you are satisfied with your insights and the data reliability.

For example, when you have validated customer interest for a product, follow up with experiments that validate the willingness of channel partners to stock and promote your product.

Execute.
When you are satisfied with the quality of your insights and the reliability of the data, you may directly start executing based on your findings.

For example, when you have learned and validated exactly what it takes to get channel partners interested in reselling your value proposition, start scaling up sales efforts by hiring salespeople or designing dedicated marketing material.

You experimented and learned. Now what?

How Quickly Are You Learning?

The only thing standing between you and finding out what customers and partners really want is the consistency and speed with which you and your team can propel yourself through the design/build, measure, learn cycle. This is called cycle time.

The speed at which you learn is crucial, especially during the early phases of value proposition design. When you start out, uncertainty is at its maximum. You don't know if customers care about the jobs, pains, and gains you intend to address, let alone if they're interested in your value proposition.

Therefore, it is critical that your early experiments be extremely fast and produce a maximum of learning so you can adapt rapidly. This is why writing a business plan or conducting a large third-party market study is the wrong thing to start with, although it can make sense later in the process.

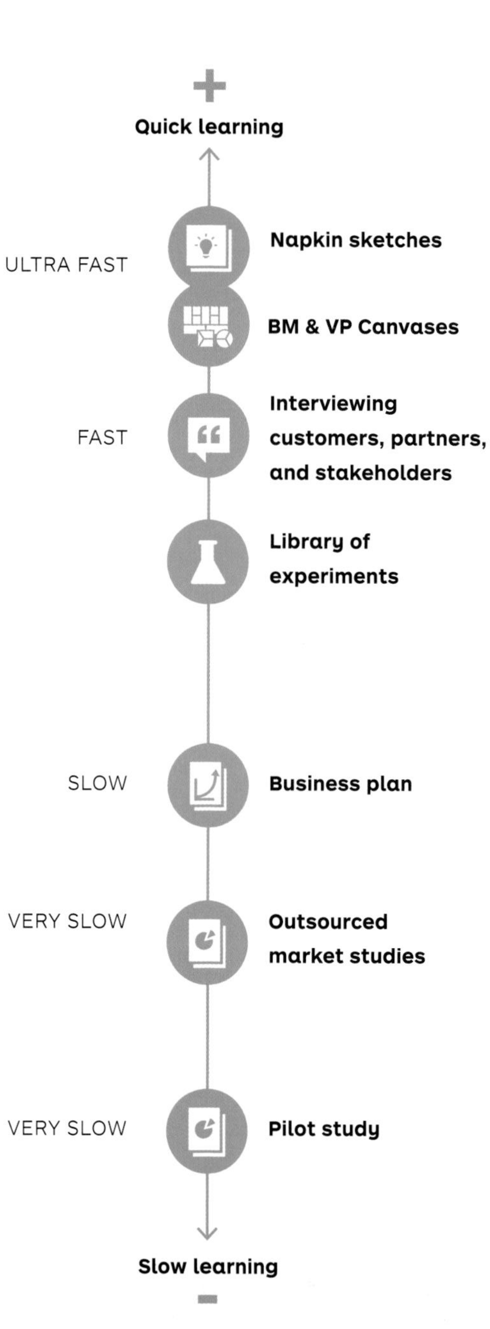

Learning Instruments

Quickly shape your ideas to share, challenge, or iterate them and to generate hypotheses to test.

Quickly gain first market insights. Keep the effort in-house so learnings remain fresh and relevant and so you can move fast and act upon insights.

Use the whole range of experiments from the experiment library ➔ p. 214. Start with quick ones when uncertainty is high. Continue with more reliable, slower ones, when you have evidence about the right direction.

Business plans are more refined documents and usually more static. Write one only when you have clear evidence and are approaching the execution phase.

Market studies are often costly and slow. They are not an optimal search tool because they don't allow you to adapt to circumstances rapidly. They make most sense in the context of incremental changes to a value proposition.

A pilot study is often the default way to test an idea inside a corporation. However, they should be preceded by quicker and cheaper learning tools, because most pilots are based on relatively refined value propositions that involve substantial time and cost.

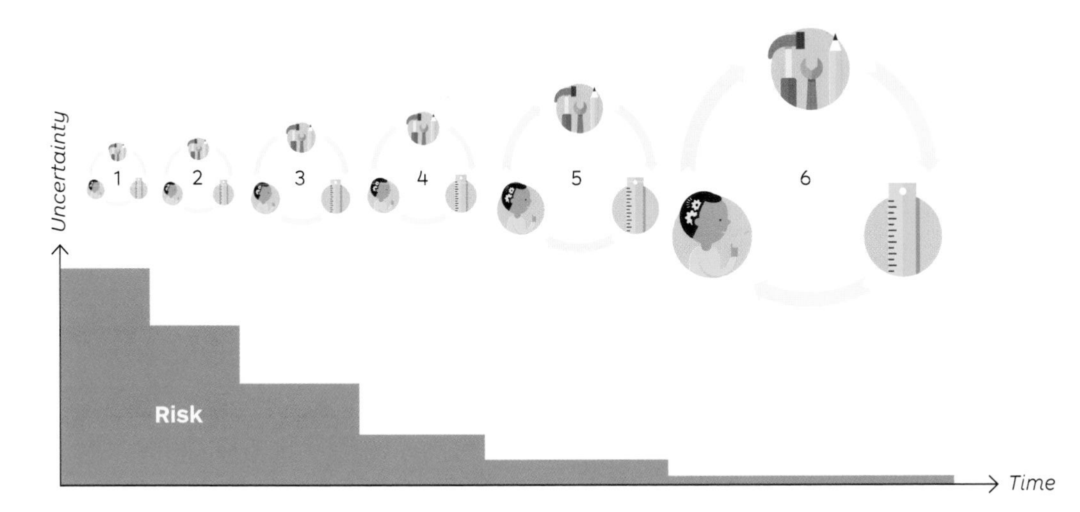

Six rapid iteration cycles based on quick experiments produce more learning than three long iteration cycles based on slower experiments. The faster approach will produce knowledge more quickly and thus reduce risk and uncertainty more substantially than the latter.

The faster you iterate, the more you learn and the faster you succeed.

Don't waste your time!
Imagine spending a week, a month, or more on refining and perfecting your idea. Imagine spending all that time thinking hard about what you'd need to do to produce great growth numbers only to find out that your customers and partners don't really care. That's wasted time!

Five Data Traps to Avoid

Avoid failure by thinking critically about your data. Experiments produce valuable evidence that can be used to reduce risk and uncertainty, but they can't predict future success with 100 percent accuracy. Also, you might simply draw the wrong conclusions from your data. Avoid the following five traps to ensure you successfully test your ideas.

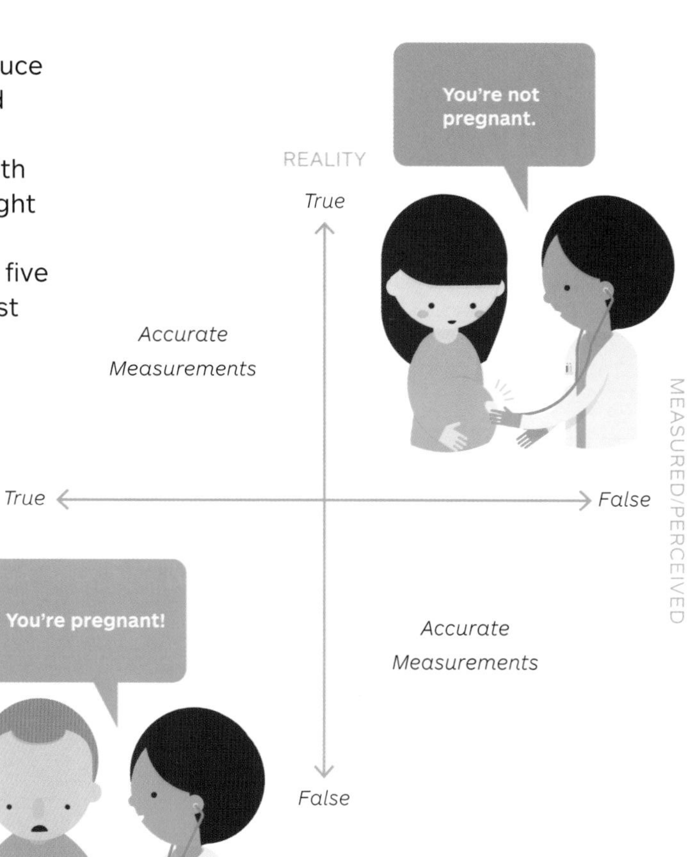

False-Positive Trap

Risk: Seeing things that are not there.
Occurs: When your testing data mislead you to conclude, for example, that your customer has a pain when in fact it is not true.

Tips

- Test the circle before you test the square. Understand what's relevant to customers to avoid being misled by positive signals for irrelevant value propositions.
- Design different experiments for the same hypothesis before making important decisions.

False-Negative Trap

Risk: Not seeing things that are there.
Occurs: When your experiment fails to detect, for example, a customer job it was designed to unearth.

Tips

Make sure your test is adequate. Dropbox, a file hosting service, initially tested customer interest with Google AdWords. They invalidated their hypotheses, because the ads didn't perform. But the reason people didn't search was because it was a new market, not because there was a lack of interest.

The "Local Maximum" Trap

Risk: Missing out on the real potential.
Occurs: When you conduct experiments that optimize around a local maximum while ignoring the larger opportunity. For example, positive testing feedback might result in you sticking with a much less profitable model when a more profitable one exists.

Tip
Focus on learning rather than optimizing. Don't hesitate to go back to designing better alternatives if the testing data are positive but the numbers feel like they should be better (e.g., larger market, more revenues, better profit).

The "Exhausted Maximum" Trap

Risk: Overlooking limitations (e.g., of a market).
Occurs: When you think an opportunity is larger than it is in reality. For example, when you think you are testing with a sample of a large population but the sample is actually the entire population.
Tip
Design tests that prove the potential beyond the immediately addressed test subjects.

The Wrong Data Trap

Risk: Searching in the wrong place.
Occurs: When you abandon an opportunity because you are looking at the wrong data. For example, you might drop an idea because the customers you are testing with are not interested and you don't realize that there are people who are interested.
Tip
Go back to designing other alternatives before you give up.

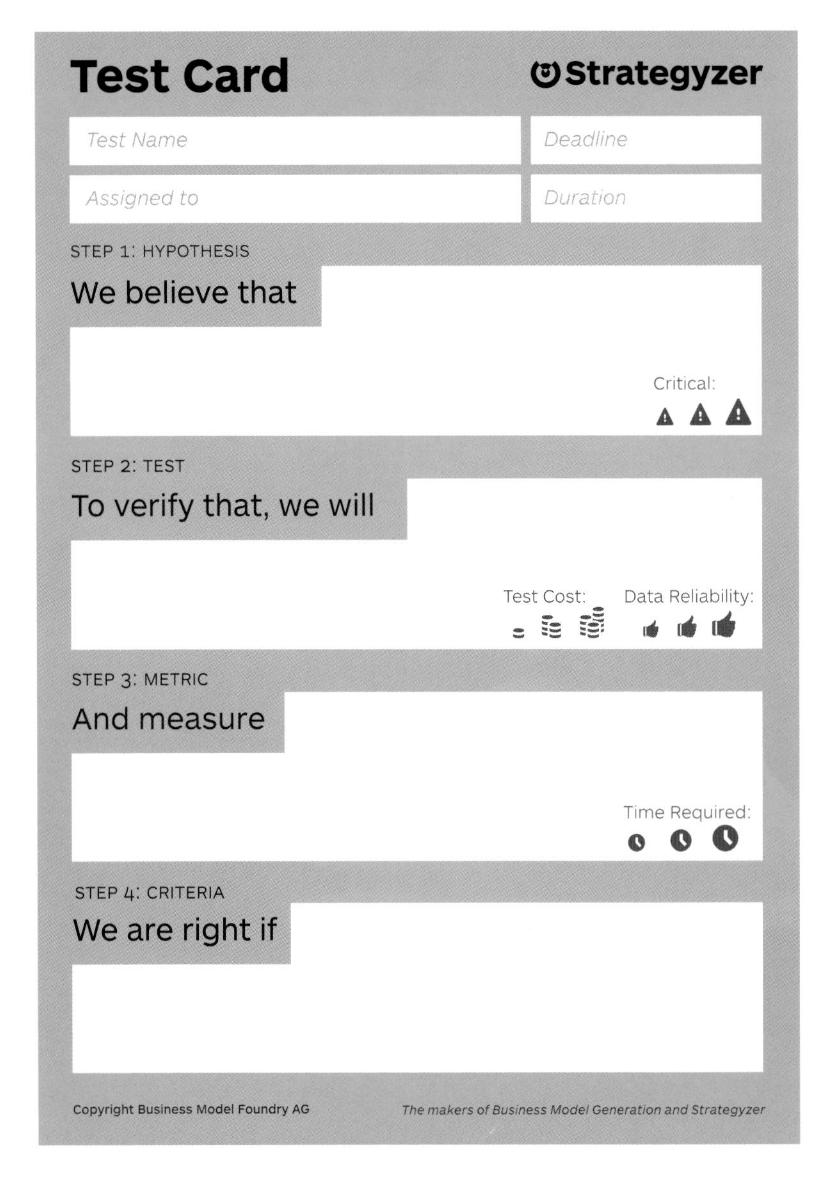

Download the Test Card

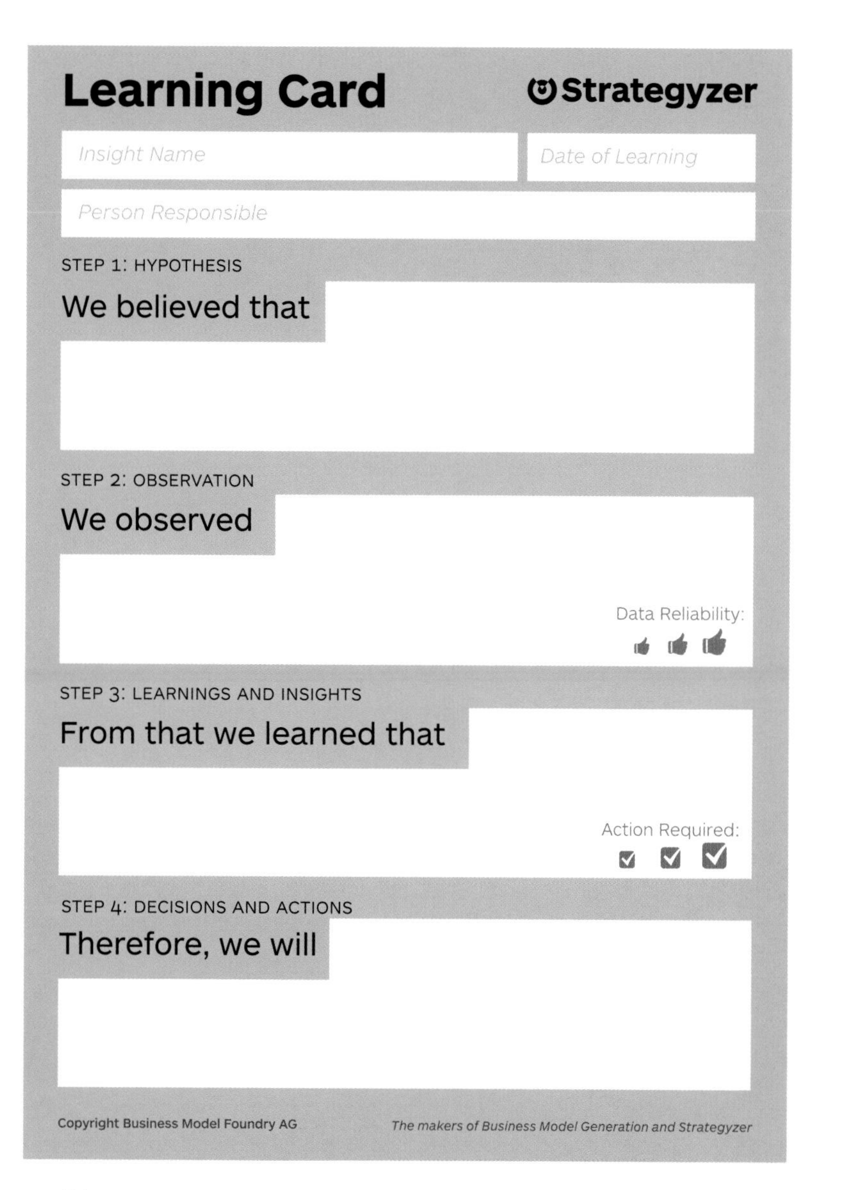

Download the Learning Card

3.3

Experiment Library

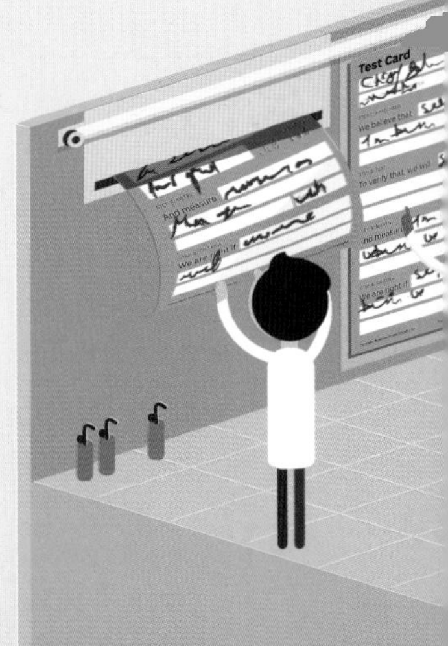

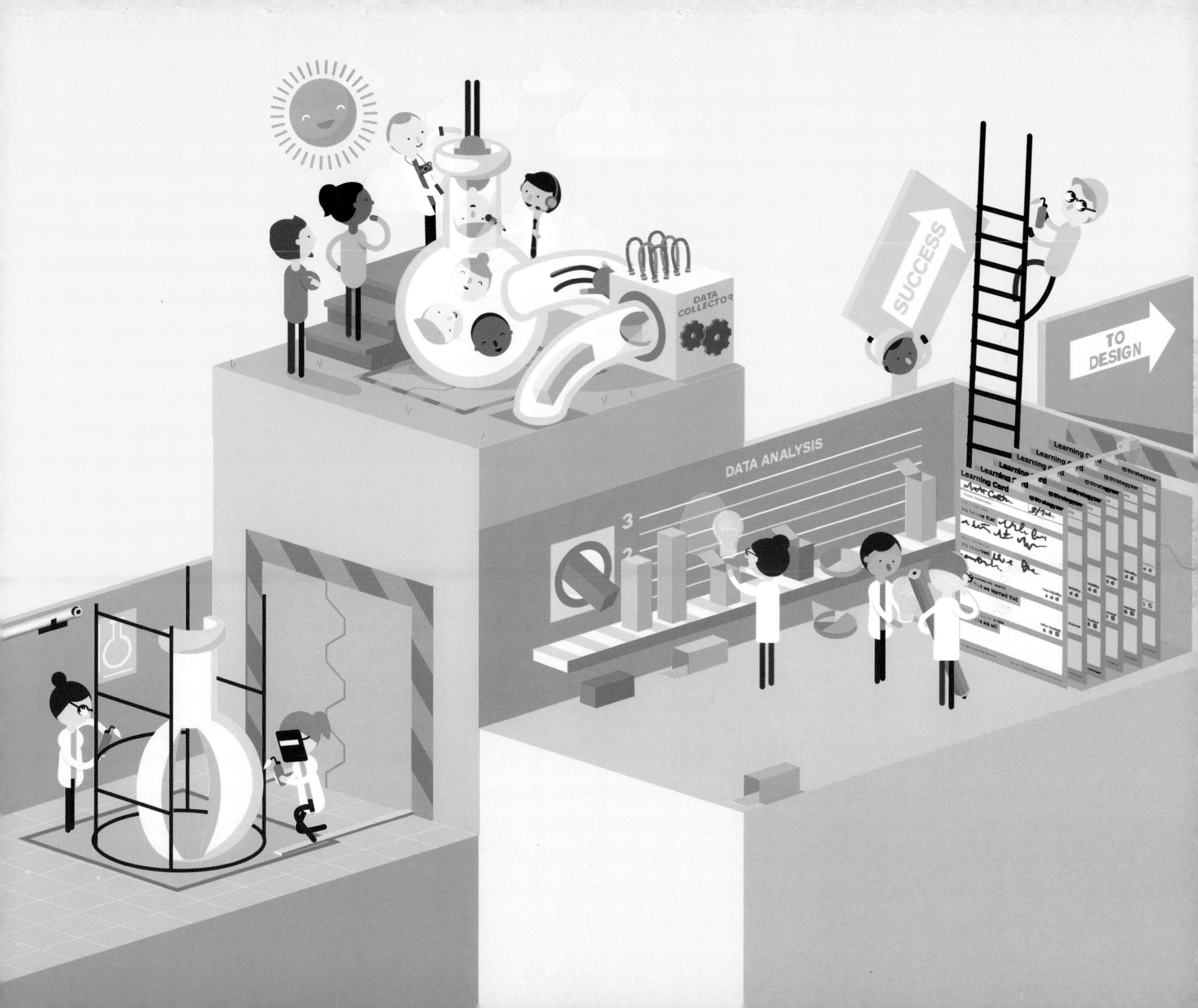
DATA COLLECTOR
SUCCESS
TO DESIGN
DATA ANALYSIS
3
2
Learning Card
Strategyzer
We believe that
We observed
From that we learned that
Therefore we will

Choose a Mix of Experiments

Every experiment has strengths and weaknesses. Some are quick and cheap but produce less reliable evidence. Some produce more reliable evidence but require more time and money to execute.

Consider cost, data reliability, and time required when you design your mix of experiments. As a rule of thumb, start cheap when uncertainty is high and increase your spending on experiments with increasing certainty.

DEF·I·NI·TION

Experiment

A procedure to validate or invalidate a value proposition or business model hypothesis that produces evidence.

Select a series of tests by drawing from our experiment library or by using your imagination to invent new experiments. Keep two things in mind when you compose your mix:

What customers say and do are two different things.
Use experiments that provide verbal evidence from customers as a starting point. Get customers to perform actions and engage them (e.g., interact with a prototype) to produce stronger evidence based on what they do, not what they say.

Customers behave differently when you are there or when you are not.
During direct personal contact with customers, you can learn why they do or say something and get their input on how to improve your value proposition. However, your presence might lead them to behave differently than if you weren't there.

In an indirect observation of customers (on the web, for example) you are closer to a real-life situation that isn't biased by your interaction with customers. You can collect numerical data and track how many customers performed an action you induced.

Tip

Use these techniques to verify whether customers really mean what they say. Produce evidence that the jobs, pains, and gains they mention are real and that they are seriously interested in your products and services.

Tip

Use these techniques to understand how customers interact with your prototypes. Investments are usually higher but produce concrete and actionable feedback.

Tip

Use these techniques at the early stages of the design process, because investment is low and they produce quick insights.

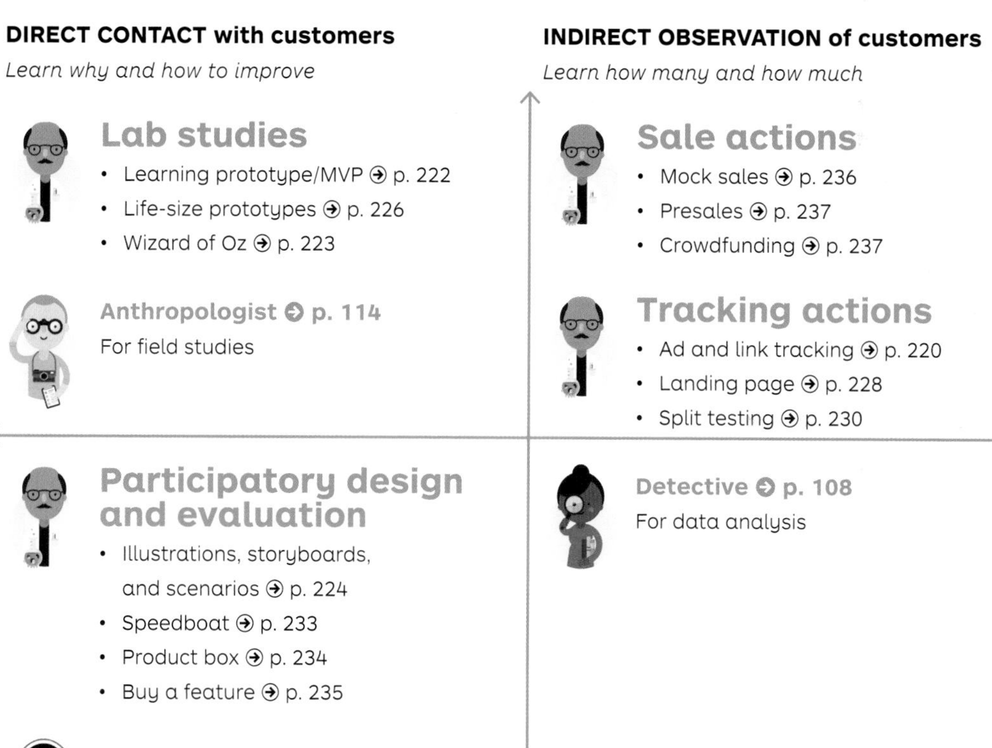

Inspired by the work in user experience by Christian Rohner (NN).

Produce Evidence with a Call to Action

Use experiments to test if customers are interested, what preferences they have, and if they are willing to pay for what you have to offer. Get them to perform a call to action (CTA) as much as possible in order to engage them and produce evidence of what works and what doesn't.

The more a customer (test subject) has to invest to perform a CTA, the stronger the evidence that he or she is really interested. Clicking a button, answering a survey, providing a personal e-mail, or making a prepurchase are different levels of investments. Select your experiments accordingly.

CTAs with a low level of investment are appropriate at the beginning of value proposition design. Those requiring a high level of investment make more sense later in the process.

DEF·I·NI·TION

Call to Action (CTA)

Prompts a subject to perform an action; used in an experiment in order to test one or more hypotheses.

Use experiments to test...

interest and relevance

Prove that potential customers and partners are genuinely interested and don't just tell you so. Show that your ideas are relevant enough to them to get them to perform actions that go beyond lip service (e.g., e-mail sign-ups, meetings with decision makers and budget holders, letters of intent, and more).

priorities and preferences

Show which jobs, pains, and gains your potential customers and partners value most and which ones they value least. Provide evidence that indicates which features of your value proposition they prefer. Prove what really matters to them and what doesn't.

willingness to pay

Provide evidence that potential customers are interested enough in the features of your value proposition to pay. Deliver facts that show they will put their money where their mouth is.

Ad Tracking

Use ad tracking to explore your potential customers' jobs, pains, gains, and interest—or lack of it—for a new value proposition. Ad tracking is an established technique used by advertisers to measure the effectiveness of ad spending. You can use the same technique to explore customer interest even before a value proposition exists.

Test customer interest with Google AdWords

We use Google AdWords to illustrate this technique because it's particularly well suited for testing based on its use of search terms for advertising (other services such as LinkedIn and Facebook also work well).

1. **Select search terms.**
   Select search terms that best represent what you want to test (e.g., the existence of a customer job, pain, or gain or the interest for a value proposition).
2. **Design your ad/test.**
   Design your test ad with a headline, link to a landing page, and blurb. Make sure it represents what you want to test.
3. **Launch your campaign.**
   Define a budget for your ad/testing campaign and launch it. Pay only for clicks on your ad, which represent interest.
4. **Measure clicks.**
   Learn how many people click on your ad. No clicks may indicate a lack of interest.

Google

Where to apply?

Test interest early in the process to learn about the existence of customer jobs, pains, gains, and interest for a particular value proposition.

Unique Link Tracking

Set up unique link tracking to verify potential customers' or partners' interest beyond what they might tell you in a meeting, interview, or call. It's an extremely simple way to measure genuine interest.

Where to apply?

This works anywhere but is particularly interesting in industries where building MVPs is difficult, such as in industrial goods and medical devices.

001

000

1

"Fabricate" a unique link.
Make a unique and trackable link to more detailed information about your ideas (e.g., a download, landing page) with a service such as goo.gl.

2

Pitch and track.
Explain your idea to a potential customer or partner. During or after the meeting (via e-mail), give the person the unique link and mention it points to more detailed information.

3

Learn about genuine interest.
Track if the customer used the link or not. If the link wasn't used, it may indicate lack of interest or more important jobs, pains, and gains than those that your idea addresses.

MVP Catalog

MVP stands for minimum viable product, a concept popularized by the lean start-up movement to efficiently test the interest in a product before building it entirely. Rather than coining a new term we stick to this established one and adapt it to testing value propositions.

Make it “real” with a representation of a value proposition.

Use the following techniques to make your value propositions feel real and tangible before implementing anything when you test them with potential customers and partners.

What’s an MVP in this book?

A representation or prototype of a value proposition designed specifically to test the validity of one or more hypotheses/ assumptions.

The goal is to do so as quickly, cheaply, and efficiently as possible. MVPs are mainly used to explore potential customer and partner interest.

Tip

Start cheaply, even in large companies with big budgets. For example, use your smartphone to make and test reactions to a video before you bring in a video crew to “professionalize” videos and expand testing.

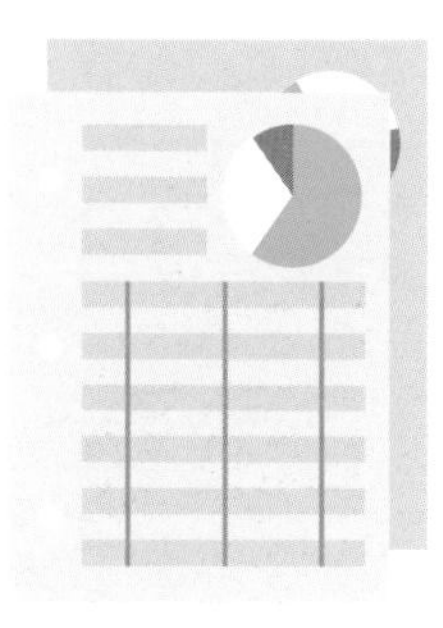

Data Sheet

Specs of your imagined value proposition

Requirements:

Word processor

Brochure

Mocked-up brochure of your imagined value proposition

Requirements:

Word processor and design skills

Storyboard

Illustration of a customer scenario showcasing your imagined value proposition

Requirements:

Sketch artist

Learn with Functional MVPs

Use prototypes designed specifically to learn from experiments with potential customers and partners.

Landing Page
Website outlining your imagined value proposition (mostly with a CTA).
Requirements:
Web designer

Product Box
Prototype packaging of your imagined value proposition
Requirements:
Packaging designer and prototype implementation

Video
Video showcasing your imagined value proposition or explaining how it works
Requirements:
Video crew

Learning Prototype
Functioning prototype of your value proposition with the most basic feature set required for learning
Requirements:
Product development

Wizard of Oz
Set up a front that looks like a real working value proposition and manually carry out the tasks of a normally automated product or service
Requirements:
Getting your hands dirty

Illustrations, Storyboards, and Scenarios

Share illustrations, storyboards, and scenarios related to your value proposition ideas with your potential customers to learn what really matters to them. These types of Illustrations are quick and cheap to produce and make even the most complex value propositions tangible.

Tips

- In a business-to-business (B2B) context think of value propositions for each important customer segment, such as users, budget owners, decision makers, and so on.
- For existing organizations, make sure to include customer-facing staff in the process, notably to get buy-in and gain access to customers to present the illustrations.
- Complement the illustrations with mock data sheets, brochures, or videos to make your ideas even more tangible.
- Run A/B tests with slightly different scenarios to capture which variations get most traction.
- Four or five meetings per customer segment are typically sufficient to generate meaningful feedback.
- Leverage the customer relationship and repeat the process later on with more sophisticated prototypes.

Process adapted from Christian Doll, bicdo.de.

1

Prototype alternative value propositions.
Come up with several alternative prototypes for the same customer segment. Go for diversity (i.e., 8-12 radically different value propositions) and variations (i.e., slightly different alternatives).

2

Define scenarios.
Sketch out scenarios and storyboards that describe how a customer will experience each value proposition in a real-world setting.

3

Create compelling visuals.
Use an illustrator to consolidate your sketches into compelling visuals that make the customer experience clear and tangible. Use single illustrations for each value proposition or entire story boards.

Questions to ask customers:

Which value propositions really create value for you?

Which ones should we keep and move forward with, and which ones should we abandon?

Dig deeper for each value proposition; pay attention to jobs, pains, and gains; and inquire:

- What is missing?
- What should be left aside?
- What should be added?
- What should be reduced?
- Always ask why to capture qualitative feedback.

4

Test with customers.
Meet customers and present the different illustrations, scenarios, and storyboards to start a conversation, provoke reactions, and learn what matters to them. Get customers to rank value propositions from most valuable to least helpful.

5

Debrief and adapt.
Use the insights from your meetings with customers. Decide which value propositions you will continue exploring, which ones you will abandon, and which ones you will adapt.

Life-Size Experiments

Get your customers to interact with life-size prototypes and real-world replicas of service experiences. Stick to the principles of rapid, quick, and low-cost prototyping to gather customer insights despite the more sophisticated set-up. Add a CTA to validate interest.

Concept Cars and Life-Size Prototypes

These are cars made to showcase new designs and technologies. Their purpose is get reactions from customers rather than to go into mass production directly.

Lit Motors used lean start-up principles to prototype and test a fully electric, gyroscopically stabilized two-wheel drive with customers. Because this type of vehicle represents a completely new concept, it was essential for Lit Motors to understand customer perception and acceptance from the very beginning.

In addition, Lit Motors added a CTA to validate customer interest beyond initial interactions with the prototype. Customers can prereserve a vehicle with a deposit ranging from $250 to $10,000. Deposits go into a holding account until vehicles are ready, with higher deposits moving customers to the front of the waiting list.

Prototype Spaces

These are spaces to cocreate products and service experiences with customers and/or observe their behavior to gain new insights. Invite potential customers to create their own perfect experience. Include industry experts to help build and test new concepts and ideas.

Hotel chain Marriott built a prototyping space in its headquarters' basement called the Underground. Guests and experts are invited to create the hotel experience of the future by cocreating hotel rooms and other spaces. Guests are invited to add furniture, electricity outlets, electronic gadgets, and more, to hotel room replicas that can easily be reconfigured.

Tips

- Make sure you validate life-size prototypes and service experiences with a CTA. Customers will always be tempted to create the perfect experience in a prototype setting, whereas they might not be willing to pay for it in real life.
- Use quicker and cheaper validation methods before you draw on life-size prototypes and real-world replicas of service experiences.
- Don't let the costs for this type of prototyping get out of hand. Stick to the principles of rapid, quick, and low-cost prototyping as much as possible, while offering a close-to-life experience to test subjects.

Landing Page

The typical landing page MVP is a single web page or simple website that describes a value proposition or some aspects of it. The website visitor is invited to perform a CTA that allows the tester to validate one or more hypotheses. The main learning instrument of a landing page MVP is the conversion rate from the number of people visiting the site to visitors performing the CTA (e.g., e-mail sign-up, simulated purchase).

> "The goal of a landing page MVP is to validate one or more hypotheses, not to collect e-mails or sell, which is a nice by-product of the experiment."

When?

Test early to learn if the jobs, pains, and gains you intend to address and/or your value proposition are sufficiently important to your customer for them to perform an action.

Variations

Combine with split testing to investigate preferences or alternatives that work better than others. Measure click activity with so-called heat maps to learn where visitors click on your page.

Use your value map to craft the headline and text that describes your value proposition on the landing page.

Test Card Strategyzer

Test Name | Deadline
Assigned to | Duration

STEP 1: HYPOTHESIS
We believe that
Critical:

STEP 2: TEST
To verify that, we will
Test Cost: Data Reliability:

STEP 3: METRIC
And measure
Time Required:

STEP 4: CRITERIA
We are right if

Copyright Business Model Foundry AG — The makers of Business Model Generation and Strategyzer

Design your landing page, traffic generation, and CTA based on your learning goals.

Traffic
Generate traffic to your landing page MVP with ads, social media, or your existing channels. Make sure you address the target customers you want to learn about, not just anybody.

Headline
Craft a headline that speaks to your potential customers and introduces the value proposition.

Value Proposition
Use the previously described techniques to make your value proposition clear and tangible to potential customers.

Call to Action
Get website visitors to perform an action that you can learn from (e.g., e-mail sign-up, surveys, fake purchase, prepurchase). Limit your CTAs to optimize learning.

Outreach
Reach out to people who performed your CTA and investigate why they were motivated enough to perform the action. Learn about their jobs, pains, and gains. Of course, this requires collecting contact information during the CTA.

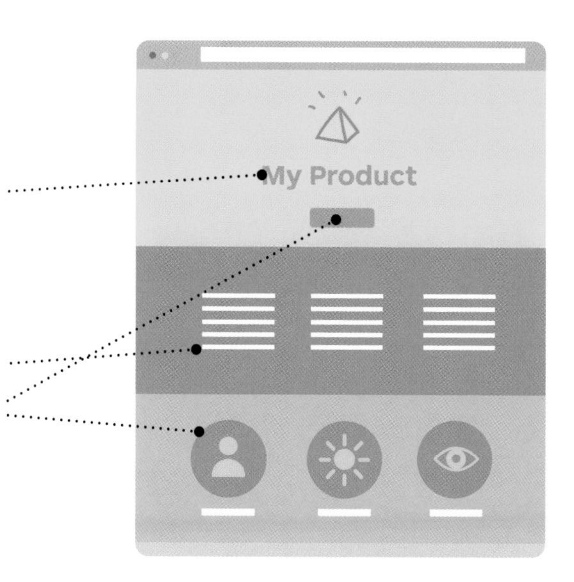

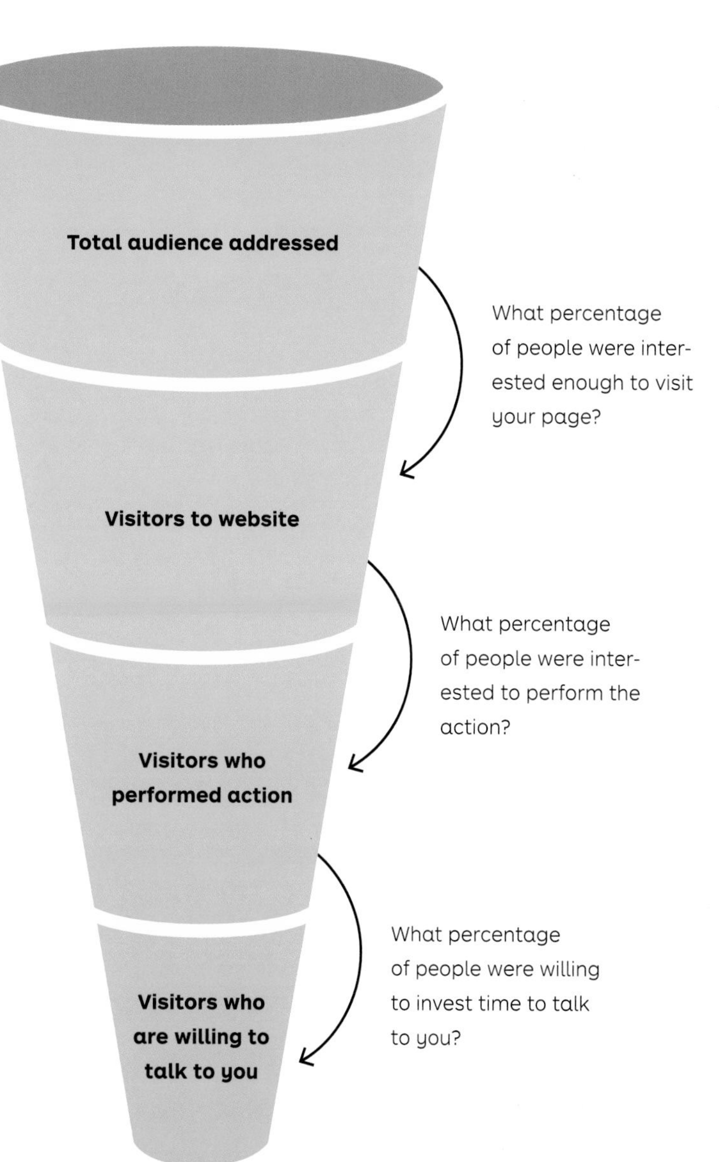

Tips

- Consider building a landing page MVP that gives the illusion that a value proposition exists even if doesn't yet. Your insights from a CTA closer to reality (e.g., simulated sales) will provide more realistic evidence than, for example, the e-mail sign-up to a planned value proposition or a prepurchase of it.
- Be transparent with your test subjects after a concluded experiment if you, for instance, "fake" the existence of a value proposition. Consider offering them a reward for participating in the experiment.
- A landing page MVP can be set up as a stand-alone web page or within an existing website.

Split Testing

Split testing, also known as A/B testing, is a technique to compare the performance of two or more options. In this book we apply the technique to compare the performance of alternative value propositions with customers or to learn more about jobs, pains, and gains.

Control

Send the same amount of people to the different options you want to test.

Compare how each option performs regarding your CTA.

Challenge

Conducting Split Tests

The most common form of split test is to test two or more variations of a web page or a purpose-built landing page (e.g., the variations may have design tweaks or outline slightly or entirely different value propositions). This technique was popularized by companies such as Google and LinkedIn, as well as the 2008 Obama campaign. Split tests can also be conducted in the physical world. The main learning instrument is to compare if conversion rates regarding a specific call to action differ between competing alternatives.

What to Test?

Here are some elements that you can easily test with A/B testing

- Alternative features
- Pricing
- Discounts
- Copy text
- Packaging
- Website variations
- ...

Call to Action

How many of the test subjects perform the CTA?

- Purchase
- E-mail sign-up
- Click on button
- Survey
- Completion of any other task

Split testing the title of this book

For this book we performed several split tests. For example, we redirected traffic from businessmodelgeneration.com to test three different book titles. We tested the titles with more than 120,000 people over a period of 5 weeks.

There were several CTAs. The first one was to simply click on a button labeled "learn more." Then people could sign up with their e-mail for the launch of the book. In the last CTA, we asked them to fill out a survey to learn more about their jobs, pains, and gains. As a small reward we showed people a video explaining the Value Proposition Canvas.

Tips

- Test a single variation in the challenging option if you want to clearly identify what leads to a better performance.
- Use so-called multivariate testing to test several combined elements to figure out which combination creates the highest impact.
- Use Google AdWords or other options to attract test subjects.
- Make sure you reach a statistical significance of greater than 95 percent
- Use tools such as Google Website Optimizer, Optimizely, or others to easily perform split tests.

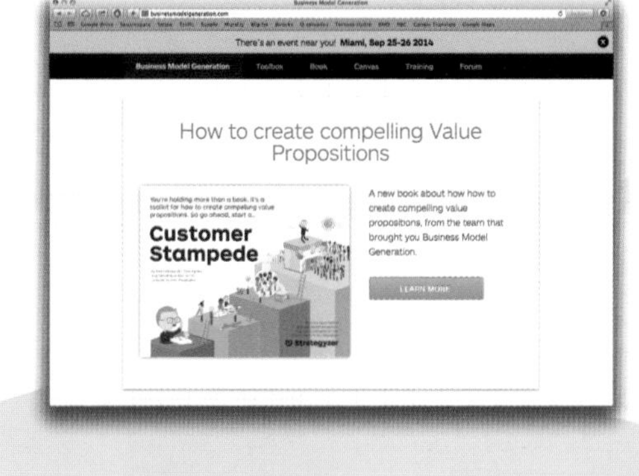

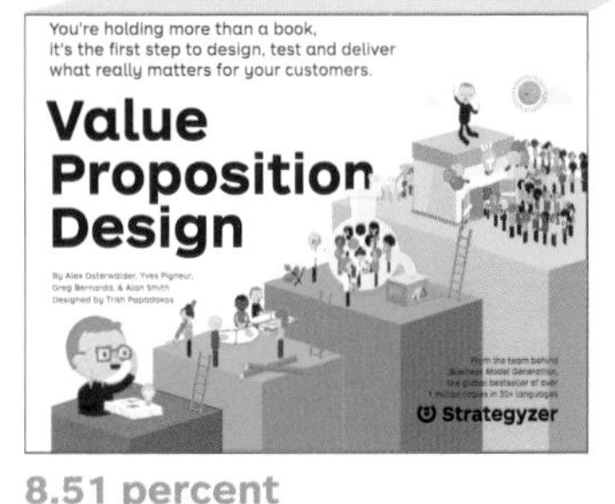

You're holding more than a book, it's a toolkit for how to create compelling value propositions. So go ahead, start a...
Customer Stampede
Strategyzer

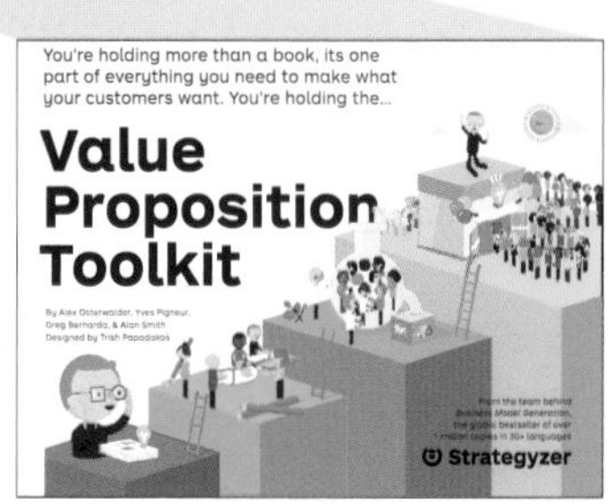

Conversion rates: 8.51 percent | **6.62 percent** | **8.21 percent**

Innovation Games®

Innovation Games is a methodology popularized by Luke Hohmann to help you design better value propositions by using collaborative play with your (potential) customers. The games can be played online or in person. We present three of them.

All three Innovation Games we present can be used in various ways. We outline three specific tasks they can help us with when it comes to the Value Proposition Canvas and related hypotheses.

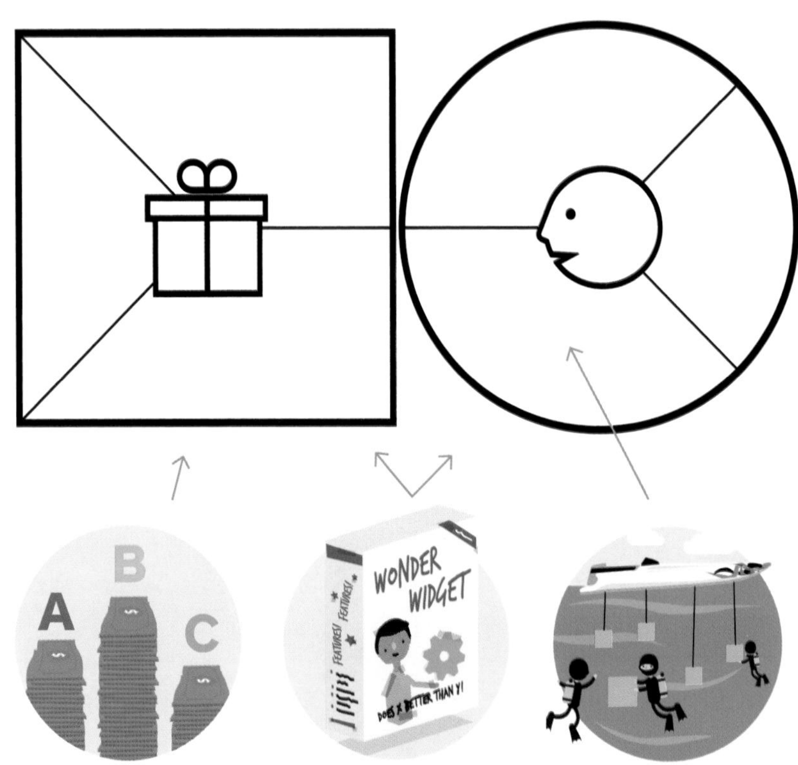

Buy a Feature
Task: Prioritize which features customers want most.

Product Box
Task: Understand your customers' jobs, pains, and gains and the value propositions they'd like.

Speed Boat
Task: Identify the most extreme pains holding customers back from completing their jobs to be done.

Hohmann, Innovation Games, 2006.

Speed Boat

This is a simple but powerful game to help you verify your understanding of customer pains. Get your customers to explicitly state the problems, obstacles, and risks that are holding them back from successfully performing their jobs to be done by using the analogy of a speed boat held back by anchors.

1

Set-up.
Prepare a large poster with a speed-boat floating at sea.

2

Identify pains.
Invite customers to identify the problems, obstacles, and risks that are preventing them from successfully performing their jobs. Each issue should go on a large sticky note. Ask them to place each sticky as anchors to the boat—the lower the anchor, the more extreme the pain.

3

Analysis.
Compare the outcomes of this exercise with your previous understanding of what is holding customers back from performing their jobs to be done.

Tips

- This exercise can be used during the design phase to identify customer pains or during testing to verify your existing understanding.
- Use a sailing boat with anchors and sails if you want to work on pains and gains simultaneously. The sails allow you to ask, "What makes the boat faster," in addition to using the anchors to symbolize what holds people back.

Product Box

In this game, you ask customers to design a product box that represents the value proposition they'd want to buy from you. You'll learn what matters to customers and which features they get excited about.

1

Design.

Invite customers to a workshop. Give them a cardboard box and ask them to literally design a product box that they would buy. The box should feature the key marketing messages, main features, and key benefits that they would expect from your value proposition.

2

Pitch.

Ask your customers to imagine they're selling your product at a tradeshow. Pretend you're a skeptical prospect and get your customer to pitch the box to you.

3

Capture.

Observe and note which messages, features, and benefits customers mention on the box and which particular aspects they highlight during the pitch. Identify their jobs, pains, and gains.

Buy a Feature

This is a sophisticated game to get customers to prioritize among a list of predefined (but not yet existing) value proposition features. Customers get a limited budget of play money to buy their preferred features, which you price based on real-world factors.

Features	Price				Total Required	Bought?
	$35	20	0	10	-5	No
	$50	5	0	0	-45	No
	$70	10	35	25	0	Yes

1

Select and price features.

Select the features for which you want to test customer preferences. Price each one based on development cost, market price, or other factors that are important to you.

2

Define the budget.

Participants buy features as a group, but each participant gets a personal budget that he or she can allocate individually. Make sure the personal budget forces participants to pool resources, and the overall budget forces them to make hard choices among the features they want.

3

Have the participants buy.

Invite participants to allocate their budget among the features they want. Instruct them to collaborate with others to get more features.

4

Analyze outcomes.

Analyze which features get most traction and are bought and which ones are not.

Mock Sales

A great way to test sincere customer interest is to set up a mock sale before your value proposition even exists. The goal is make your customers believe they are completing a real purchase. This is easily done in an online context but can also be done in a physical one.

Online

Test different levels of customer commitment with these three experiments:

Learn about customer interest by measuring how many people click on a simple "buy now" button.

Learn how pricing influences customer interest. Combine with A/B testing (see p. 230) to learn more about demand elasticity and the optimal price point.

BUY NOW ($500)

Get hard data by simulating a transaction with the customer's credit card information. This is the strongest evidence of customer demand (see tips to manage customer perception, p. 237).

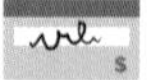

Enter credit card number

BUY NOW ($500)

Physical World

Mock sales are not limited to online. Here's what retailers do to test customer interest and pricing in the real world:

Introduce products that don't exist yet in a limited number of (mail order) catalogs.

Sell a product in one retail location only for a limited amount of time (different from a pilot, which typically covers an entire market).

Presales

The main objective of this type of presales is to explore customer interest; it is not to sell. Customers make a purchase commitment and are aware of the fact that your value proposition does not yet fully exist. In case of a lack of interest, the sale is canceled and the customer reimbursed.

Tips

Don't fear that mock sales might alienate customers and negatively affect your brand. Manage customer perception well, and mock sales can be turned into an advantage. Build on these best practices:

- Explain that you were performing a test after the customer completes the mock purchase.
- Be transparent about which information you keep or erase.
- Always erase credit card information in a fake purchase.
- Offer a reward for participating in the test (e.g. goodies, discounts).

You will turn test subjects into advocates for your brand rather than alienate them if you manage customer perception well.

Attention

Remember that successful presales are only an indicator. Ouya, an Android-based video game console, raised millions on Kickstarter but later failed to attract a large base of customers or design a scalable business model.

Online

Platforms such as Kickstarter made preselling popular. They allow you to advertise a project, and if customers like it, they can pledge money. Projects receive funds only if they reach their predefined funding goals. If you are up for building the required infrastructure you can also set up your own presales process.

Physical World

Pledges, letters of intent, and signatures, even if not legally binding, are a powerful technique to test potential customers' willingness to buy. This is also easier to apply in a B2B context.

3.4

Bringing It All Together

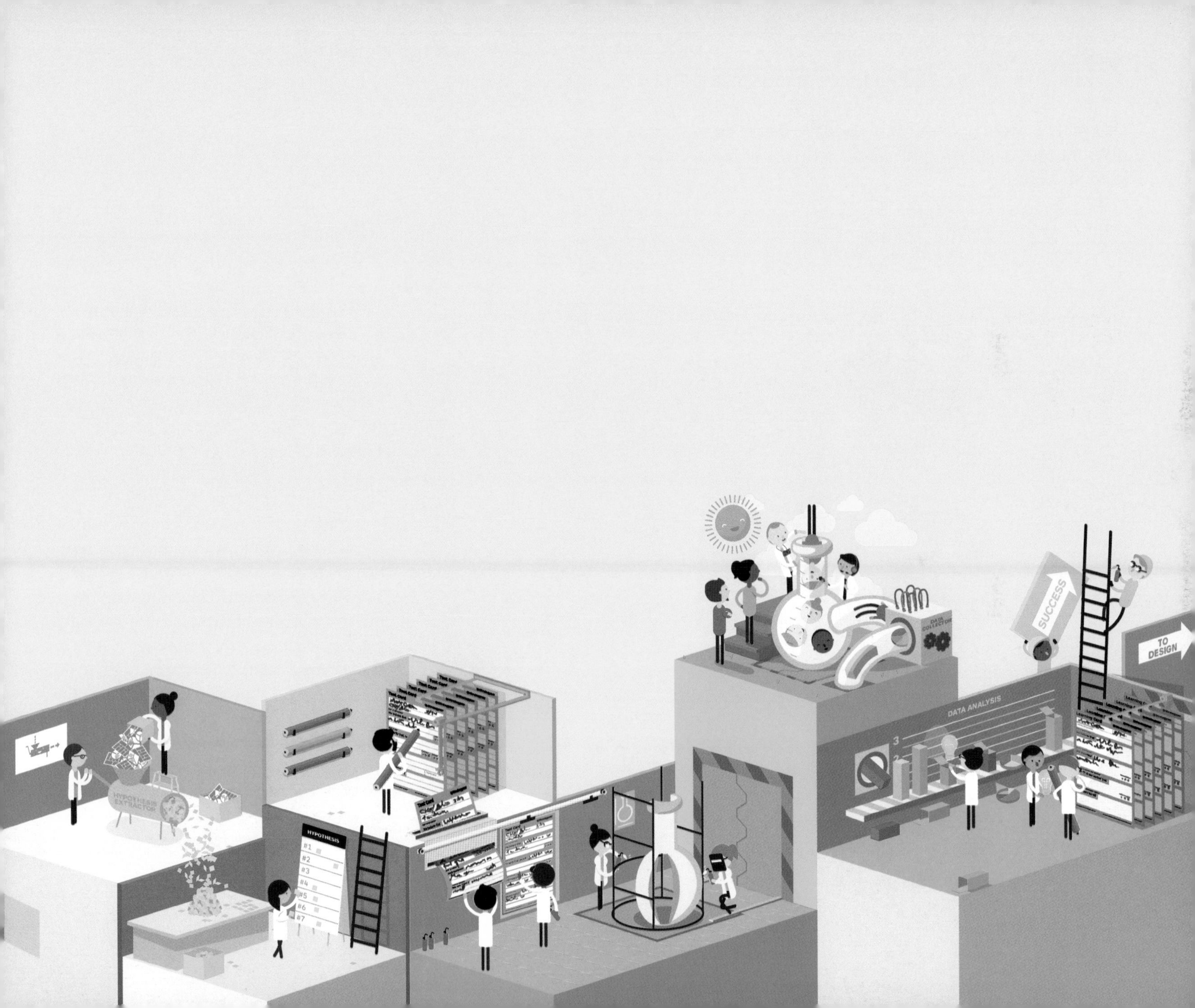
HYPOTHESIS EXTRACTOR
HYPOTHESIS
#1
#2
#3
#4
#5
#6
#7
DATA COLLECTOR
SUCCESS
DATA ANALYSIS
TO DESIGN

The Testing Process

Use all the tools you learned about to describe what you need to test and how you will do so in order to turn your idea into reality.

What to Test

With the Value Proposition and Business Model Canvases, you map out how you believe your idea could become a success. This foundation allows you to easily make the hypotheses explicit that need to be true for your idea to work. Start by testing the most important ones with a series of experiments.

How to Test

With the testing card, you describe how exactly you will verify your most important hypotheses and what you will measure. After one or more completed experiments, you use the learning card to capture your insights and indicate whether you need to learn more, iterate, pivot, or move on to test the next important hypothesis.

What's Next

Keep your eyes on the prize, and make sure you are progressing. Track whether you are advancing from your initial idea toward a profitable and scalable business via problem-solution fit, product-market fit, and business model fit.

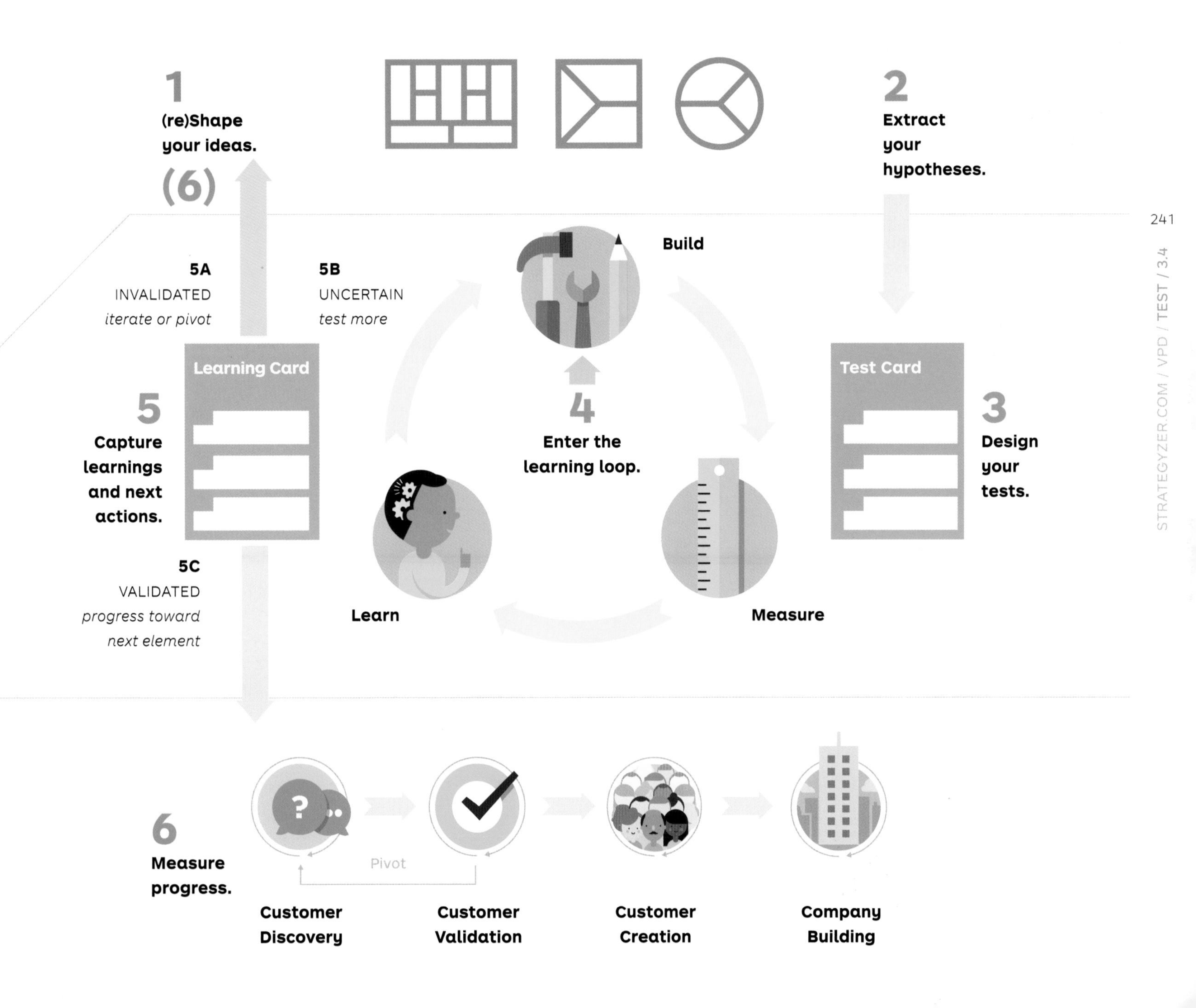
1
(re)Shape
your ideas.
(6)
2
Extract
your
hypotheses.
5A
INVALIDATED
iterate or pivot
5B
UNCERTAIN
test more
Build
Learning Card
Test Card
5
Capture
learnings
and next
actions.
4
Enter the
learning loop.
3
Design
your
tests.
5C
VALIDATED
progress toward
next element
Learn
Measure
6
Measure
progress.
Pivot
Customer
Discovery
Customer
Validation
Customer
Creation
Company
Building

Measure Your Progress

The testing process allows you to continuously reduce uncertainty and gets you closer to turning your idea into a real business. Measure your progress toward this goal by tracking the activities you've done and the results you've achieved. We designed this spread that allows you to understand if you're progressing based on Steve Blank's Investment Readiness Thermometer.

Download Progress Indicators

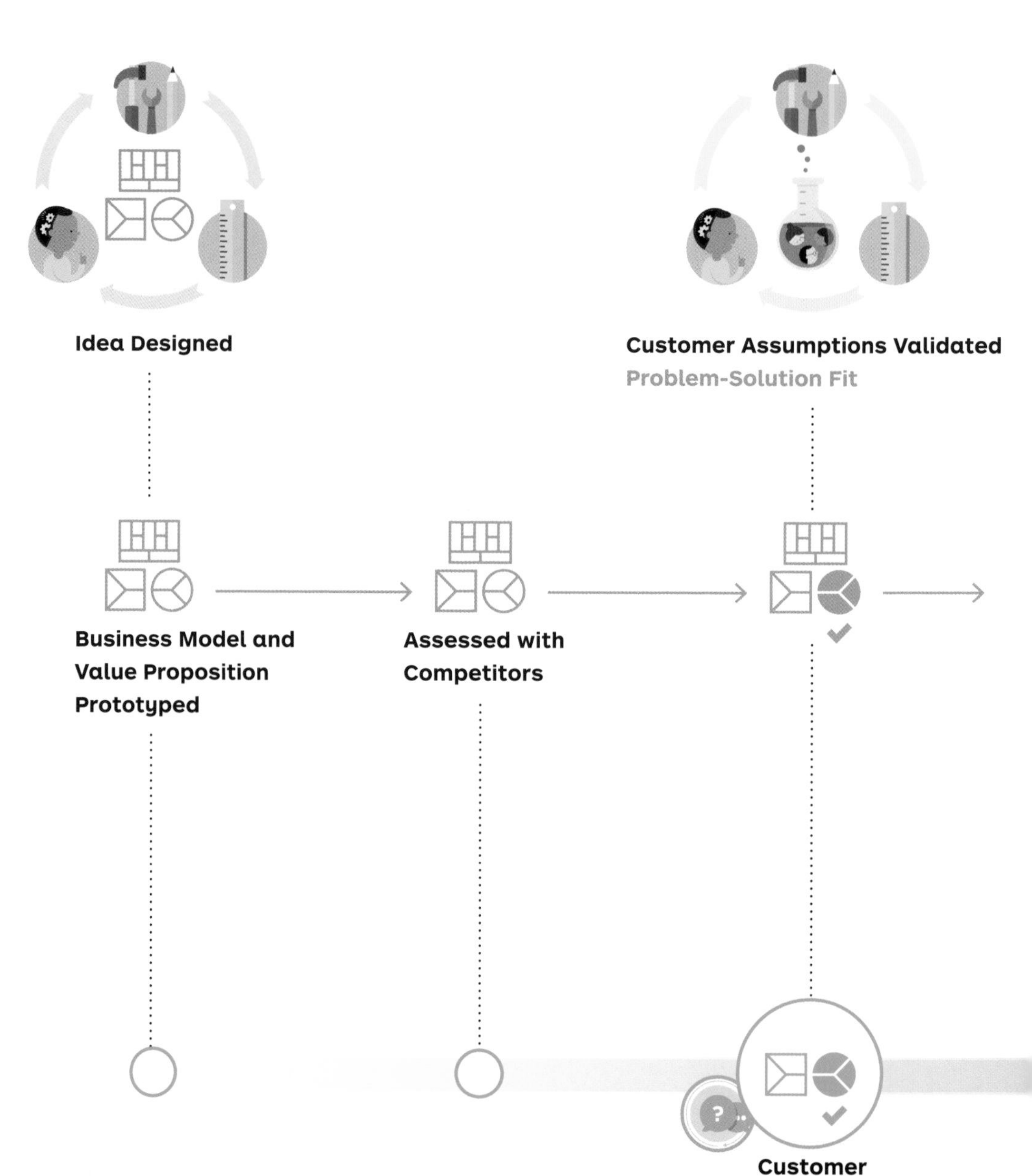

Blank, Investment Readiness Thermometer, 2013, http://steveblank.com/2013/11/25/its-time-to-play-moneyball-the-investment-readiness-level/.

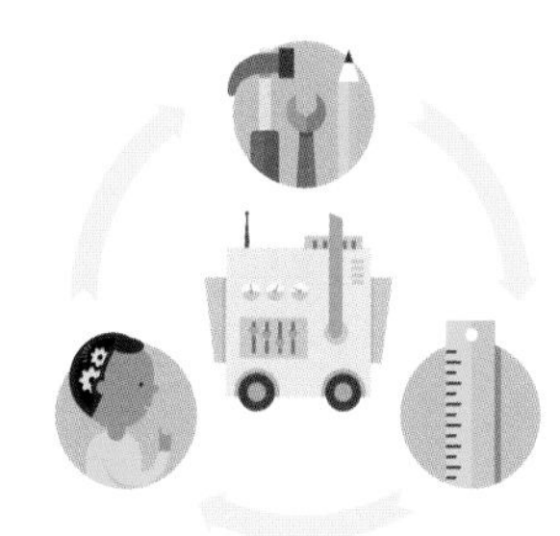

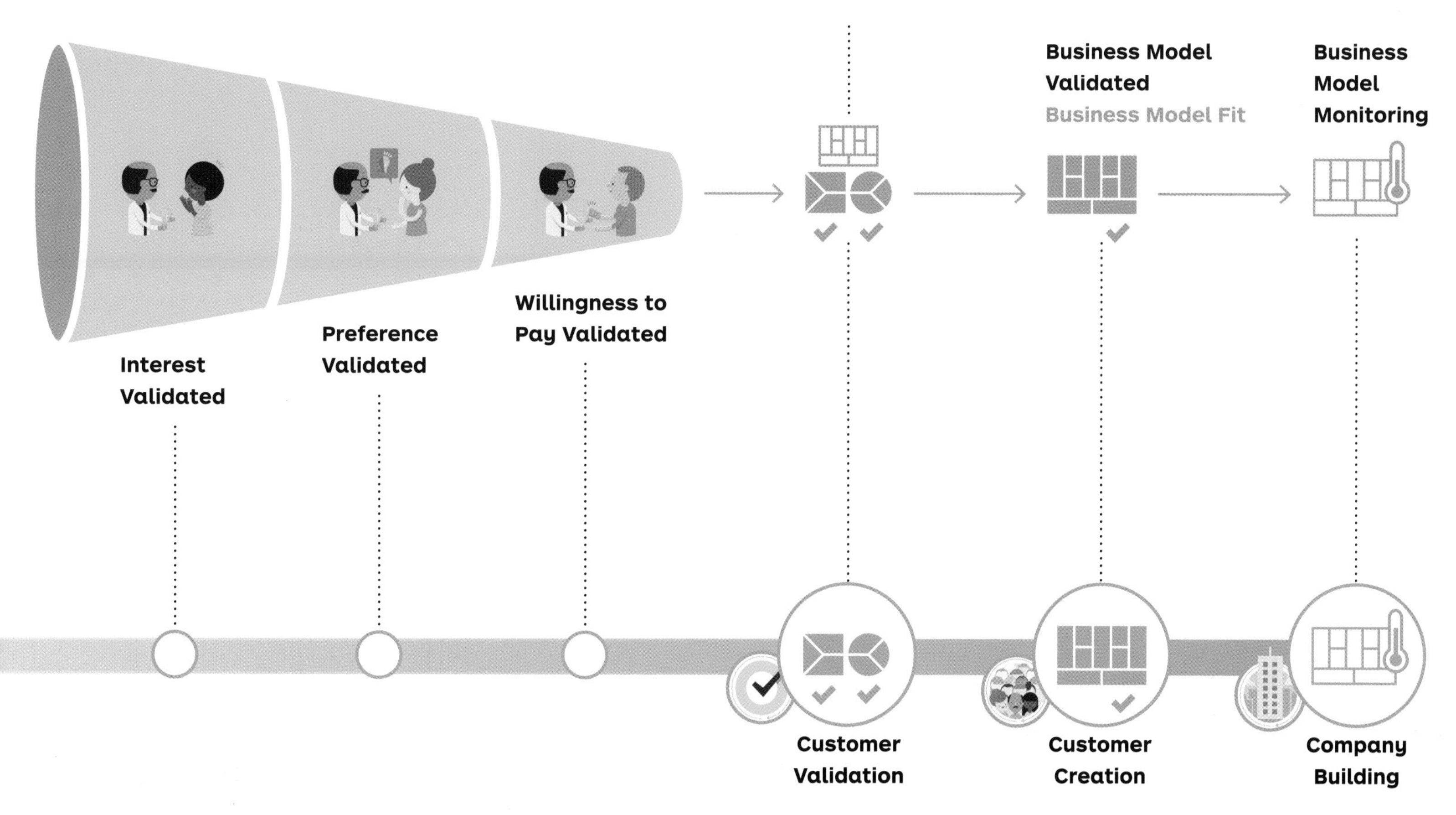
Value Proposition Validated
Product-Market Fit
Business Model Validated
Business Model Fit
Business Model Monitoring
Interest Validated
Preference Validated
Willingness to Pay Validated
Customer Validation
Customer Creation
Company Building

The Progress Board

Use the progress board to manage and monitor your tests and assess how much progress you are making toward success.

Get Progress Board poster

What did I test already?

Use the Value Proposition and Business Model Canvases to track which elements you have tested, validated, or invalidated.

What am I testing, and what did I learn?

Track the tests you are planning, building, measuring, and digesting to learn and make your insights and follow-up actions explicit.

How much progress did I make?

Keep score of how much progress you are making.

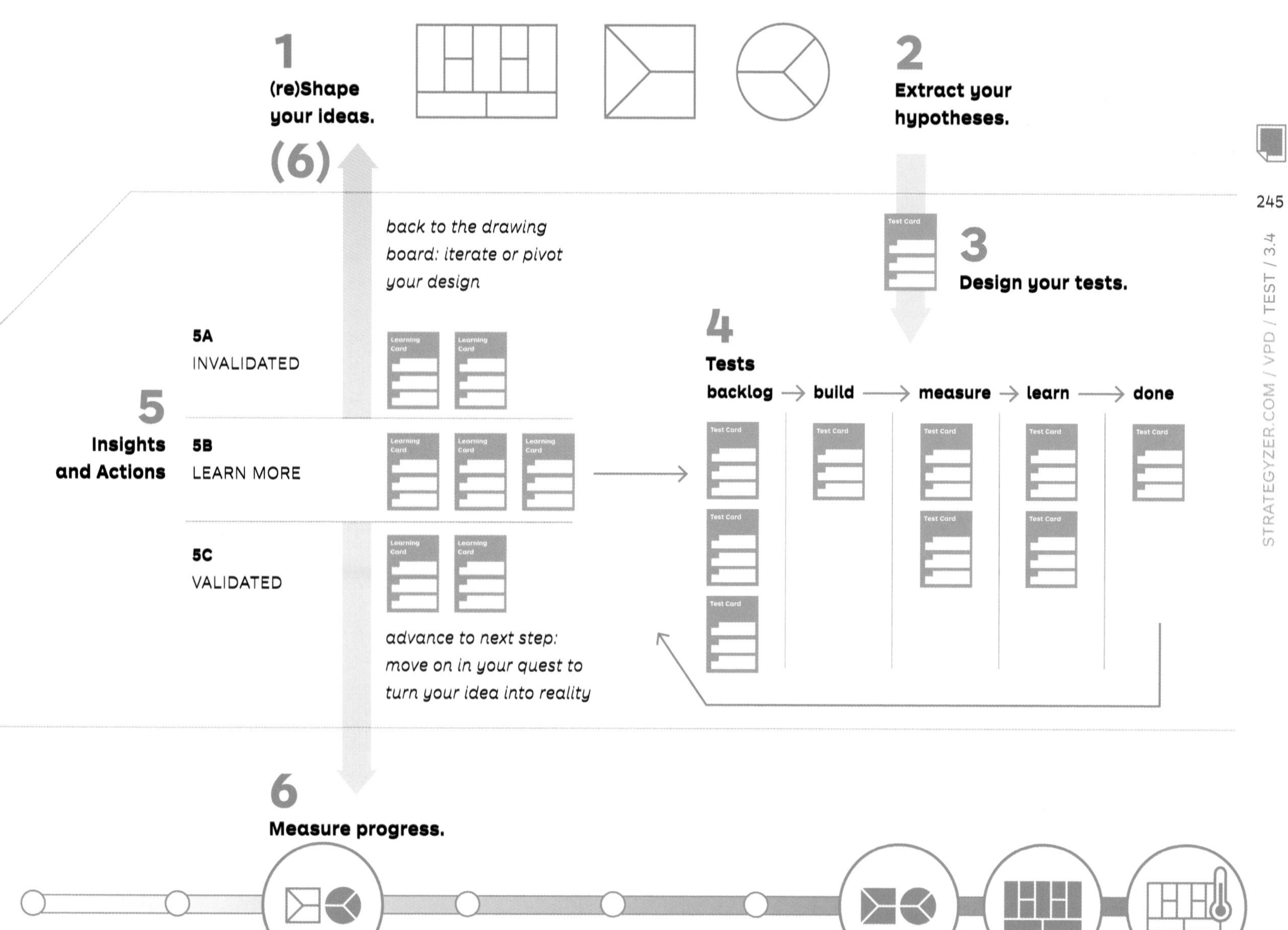
1
(re)Shape your ideas.
(6)
2
Extract your hypotheses.
3
Design your tests.
back to the drawing board: iterate or pivot your design
4
Tests
backlog
build
measure
learn
done
Test Card
5
Insights and Actions
5A
INVALIDATED
5B
LEARN MORE
5C
VALIDATED
Learning Card
advance to next step: move on in your quest to turn your idea into reality
6
Measure progress.

Owlet: Constant Progress with Systematic Design and Testing

Wireless monitoring of babies' blood oxygen, heart rate, and sleep data.*

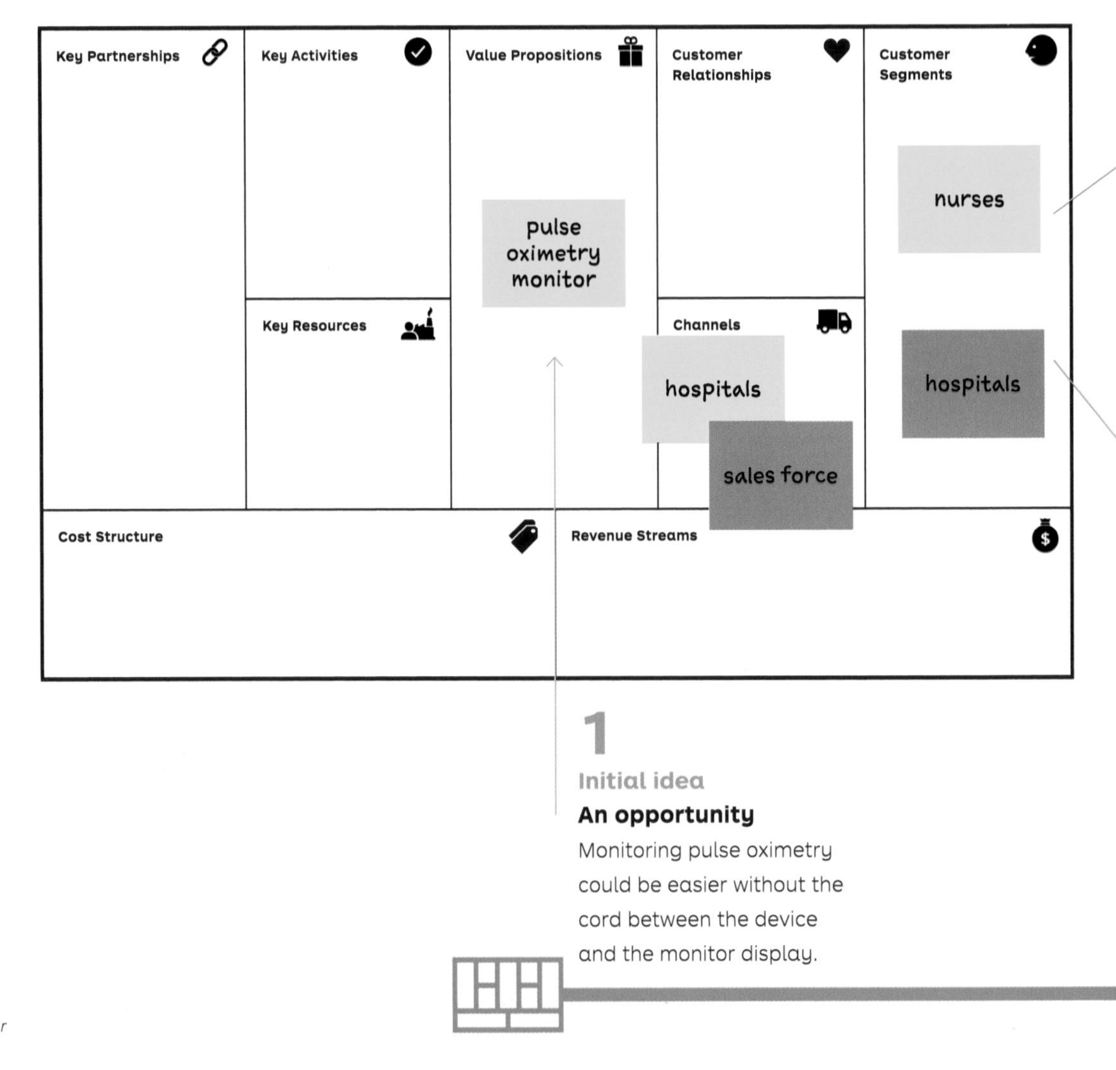

1

Initial idea

An opportunity

Monitoring pulse oximetry could be easier without the cord between the device and the monitor display.

Watch Owlet presentation online

**Case adopted in accordance with Owlet. Owlet was the winner of the 2013 International Business Model Competition.*

Test 1A: Nurse Interviews

HYPOTHESIS: Wireless pulse oximetry is more convenient

METRIC: Percentage of positive feedback

TEST: Interview nurses

DATA: Of 58 nurses interviewed, 93 percent prefer the wireless monitoring.

Validated: 1 week, $0

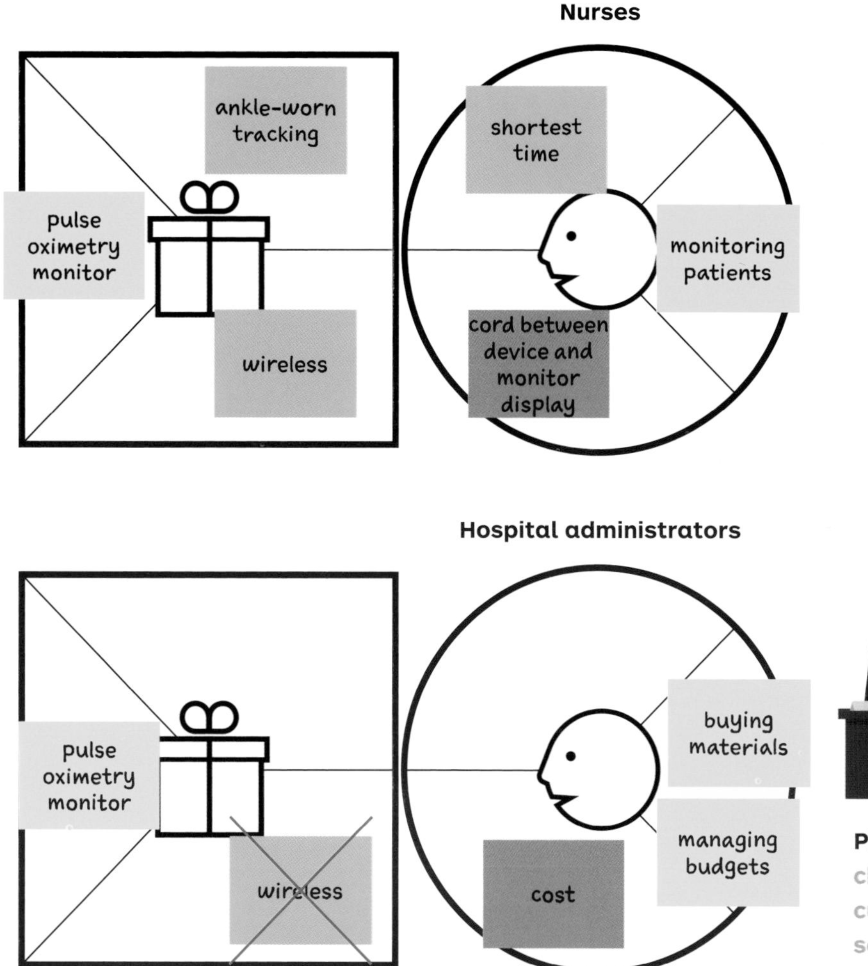

Test 1B: Hospital Administrator Interviews

HYPOTHESIS: Wireless pulse oximetry is more convenient.

METRIC: Percentage of positive feedback

TEST: Interview hospital administrators

DATA: 0 percent ready to pay more for wireless "ease of use is not a pain, if not cost-effective."

Unvalidated: 1 week, $0

Pivot: change the customer segment

DATA: *Sudden infant death syndrome (SIDS) is the leading cause of infant deaths.*

Owlet Business Model: version 2

Key Partnerships

Key Activities

Key Resources

Value Propositions: baby alarm

Customer Relationships

Channels: baby stores

Customer Segments: parents

Cost Structure

Revenue Streams: <$200 price

a first pivot after one week

Pivot:
Change the customer segment from nurses and hospitals to worried parents.

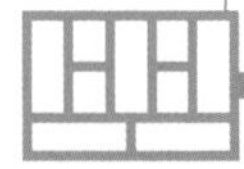

2

Iteration

Peace of mind for parents

A wireless monitor that collects the baby's heart rate, oxygen level, and sleep pattern and sends them via Bluetooth to the parents' smartphone with alerts; distributed by baby stores.

Test 2: Parent interviews

HYPOTHESIS: Parents are ready to adopt and buy a wireless baby alarm.
METRIC: Percentage of adopting parents
TEST: Interview mothers
DATA: Of 105 mothers interviewed, 96 percent adopt the wireless monitoring.
"Awesome. I want to buy now!"
Validated

Test 3: MVP Landing page

HYPOTHESIS: A smart bootie is convenient and easy to use for monitoring.
METRIC: Number of positive comments
TEST: An MVP, with a video on a website
DATA: 17,000 views, 5,500 shares of Facebook, 500 positive comments by parents, distributors, and research organizations
Validated, 2 weeks, $220

Test 4: A/B Price test

HYPOTHESIS: Rental versus sale at $200+ sale price
METRIC: Percentage for a sale price
TEST: A/B testing, 3 rounds, on the website
DATA: 1,170 people tested, $299 the best price
Validated, 8 weeks, $30

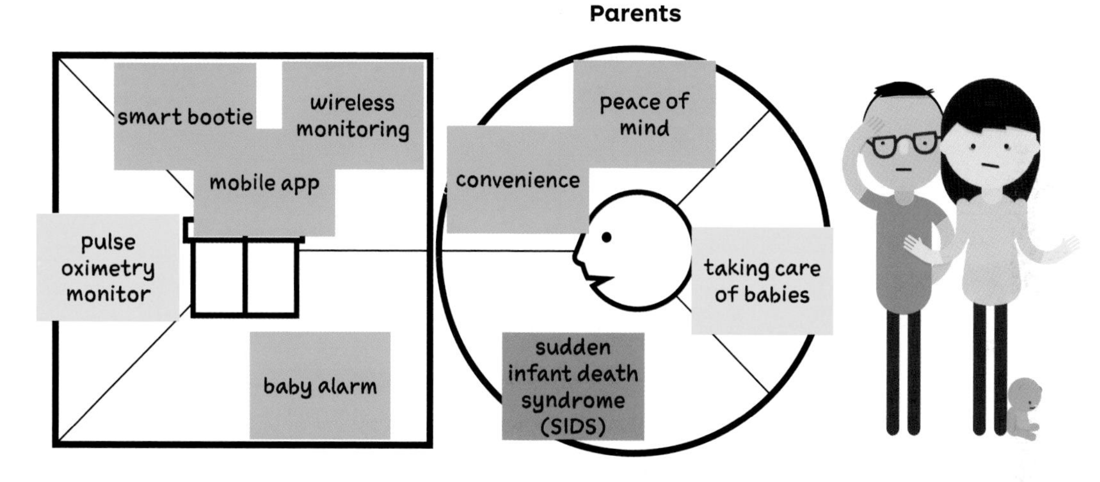

seems to be a promising business but...

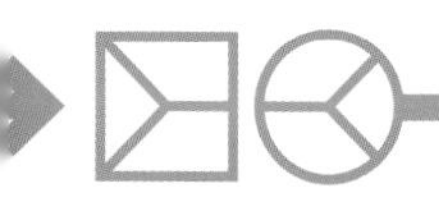

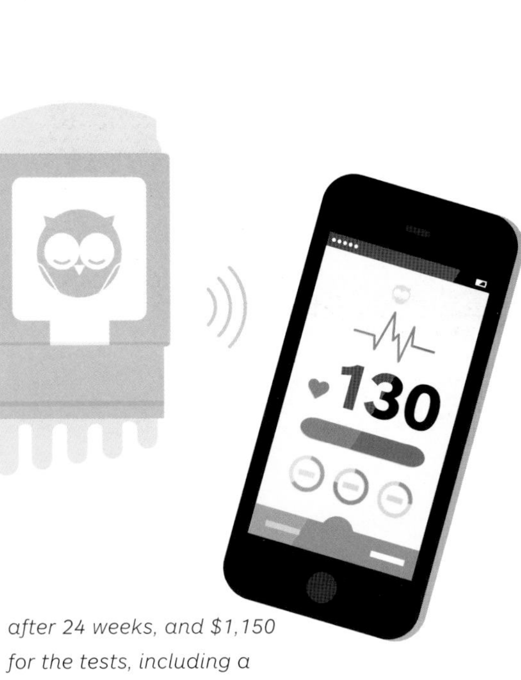

after 24 weeks, and $1,150 for the tests, including a technical proof of concept

Running lean

based on experts, a Food and Drug Administration (FDA) clearance for a baby alarm is one year, $120,000–$200,000

Owlet Business Model: version 3

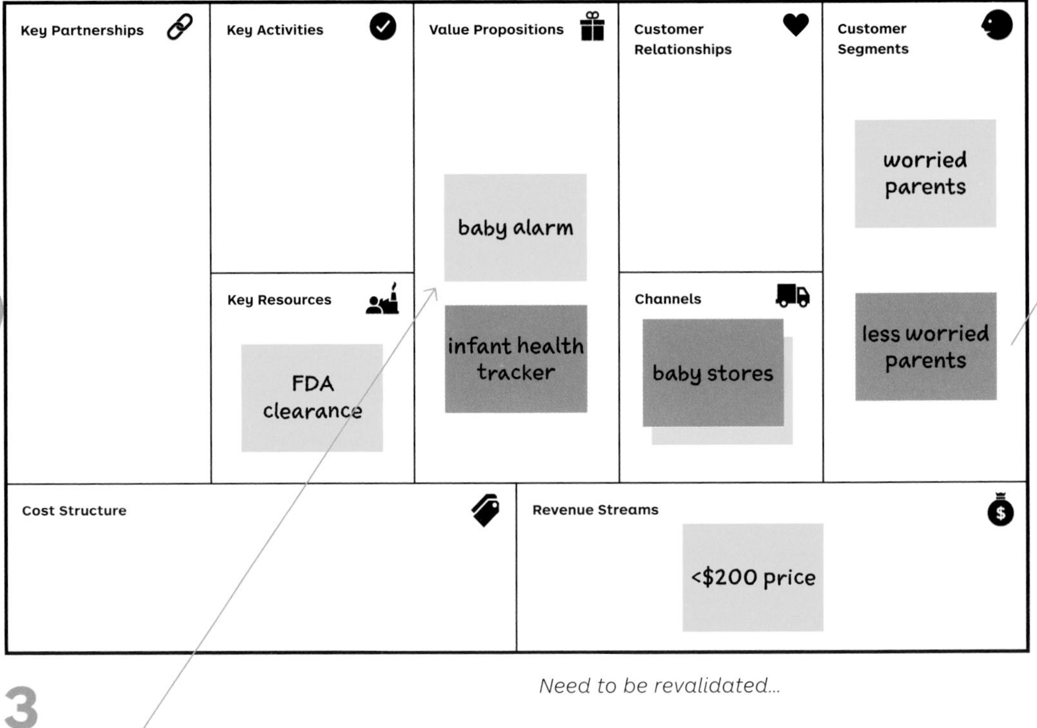

Need to be revalidated...

3

Iteration

Peace of mind, but for less worried parents

With a more minimal, less risky product, an infant health tracker (heart rate, oxygen levels, and sleep patterns), but without alarm, for another customer segment: the less worried parents.

Test 5: Interview/Proposition: "Owlet Challenge"

HYPOTHESIS: Less worried parents are ready to adopt and buy a wireless baby health tracker, without alarm.

METRIC: Percentage of parents adopting the no-alarm tracker

TEST: Interview at retail locations, having to choose between the Owlet tracker and other similar systems (video, sound, and movement)

DATA: Of 81 people interviewed, 20 percent adopted the Owlet tracker.

Validated, 3 weeks, $0

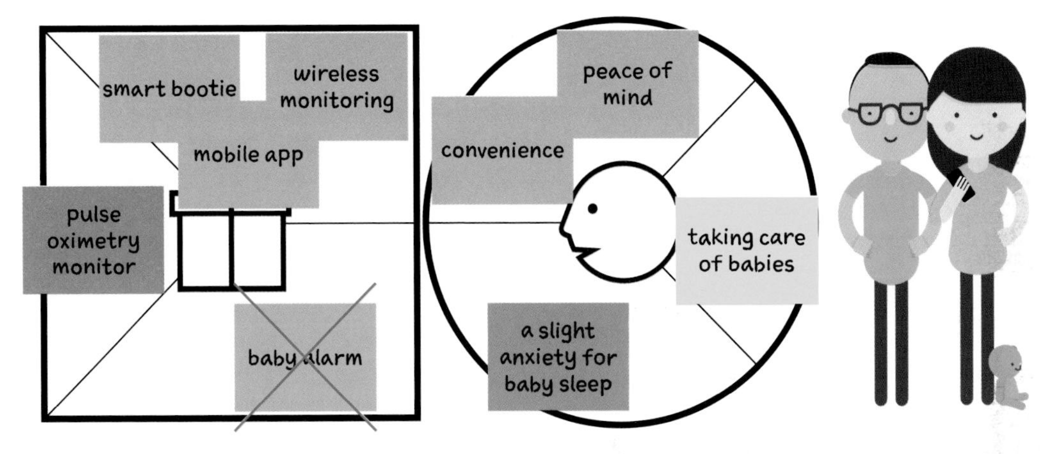

Owlet decided to start first with the baby health tracker and to come later with the baby alarm, after FDA clearance.

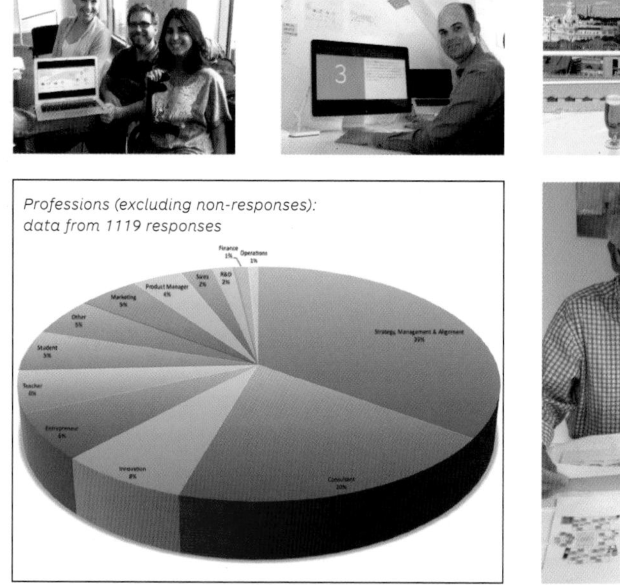

Lessons Learned

Testing Step-by-Step

Your customers are the judge, jury, and executioner of your value proposition, so get outside of the building and test your ideas with the customer development and lean start-up process. Make sure you start with quick and cheap experiments to test the assumptions underlying your ideas when uncertainty is at its maximum.

Experiment Library

What your customers say might wildly differ from what they do in reality. Go beyond talking to customers and conduct a series of experiments. Get them to perform actions that provide evidence of their interest, their preferences, and their willingness to pay.

Bringing It All Together

Launching ideas without testing is wishful thinking. Testing ideas without launching is just a pastime. Launching tested ideas can change your life as an entrepreneur. Measure your progress from idea to real business step by step.

Dashboard

Recent Campaigns

List Growth

Top 5

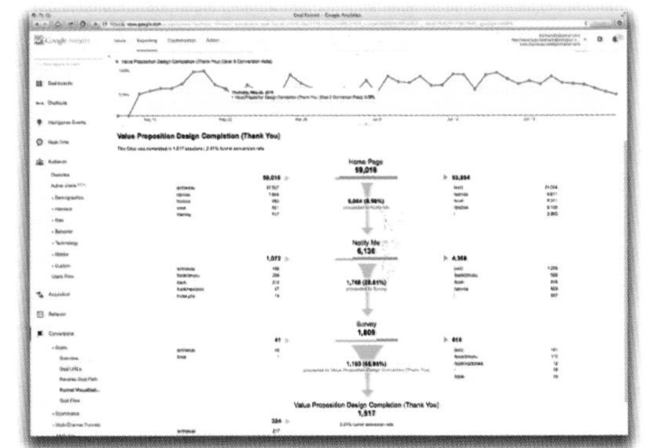

New Book Titles Test

26.71% 21.99% 23.97% 23.77% 25.09% 31.38%

863 26.69%

31.38% 26.71%

3,233

310 152

evo

lve

4

Use the Value Proposition and Business Model Canvas as a shared language to **Create Alignment** p. 260 throughout every part of your organization while it continuously evolves. Make sure you constantly **Measure and Monitor** p. 262 your value propositions and business models in order to **Improve Relentlessly** p. 264 and **Reinvent Yourself Constantly** p. 266.

Create Alignment

The Value Proposition Canvas is an excellent alignment tool. It helps you communicate to different stakeholders which customer jobs, pains, and gains you are focusing on and explains how exactly your products and services relieve pains and create gains.

Advertising

Packaging

Explainer Videos

Slide Decks

Sales Scripts

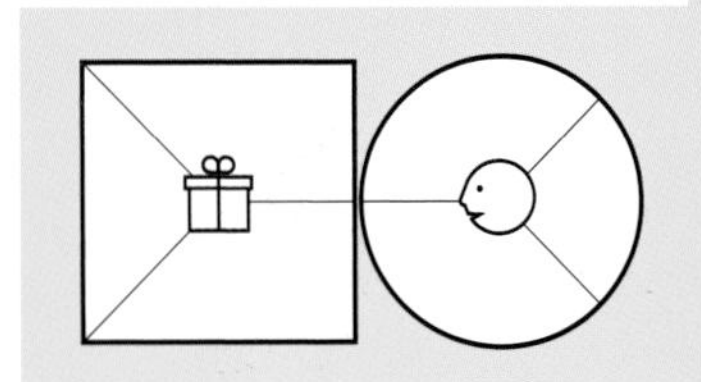

Craft aligned messages.

Align internal and external stakeholders.

Marketing

Craft marketing messages based on the jobs, pains, and gains your products and services are helping with. Align customer-facing messaging all the way from advertising to package design. Point out which pain relievers and gain creators to focus on.

(Channel) Partners

Bring (channel) partners on board, and explain your value proposition. Help them understand why customers will love your products and services by highlighting pain relievers and gain creators.

Employees

Help all employees understand which customers you are targeting and which jobs, pains, and gains you are addressing, and outline how exactly your products and services will create value for customers. Explain how the value proposition fits into the business model.

Sales

Help sales understand which segments to target and what customers' jobs, pains, and gains are. Highlight which attributes of your value proposition are most likely to sell by relieving pains and creating gains. Align sales scripts and pitch decks.

Shareholders

Explain to your shareholders how exactly you intend to create value for your customers. Clarify how the (new or improved) value proposition will bolster your business model and create a competitive advantage.

Measure and Monitor

Use the Value Proposition and Business Model Canvases to create and monitor performance indicators once your value proposition is operational in the market. Track the performance of your business model, your value proposition, and your customers' satisfaction.

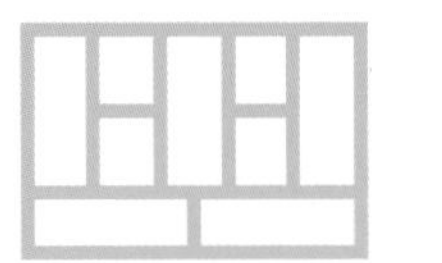

Business Model Performance

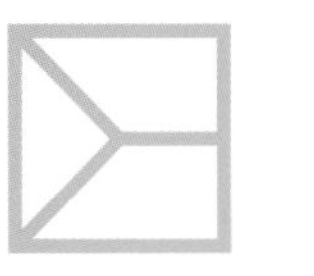

Value Proposition Performance

(Quantitative Facts)

Customer Satisfaction

(Perception)

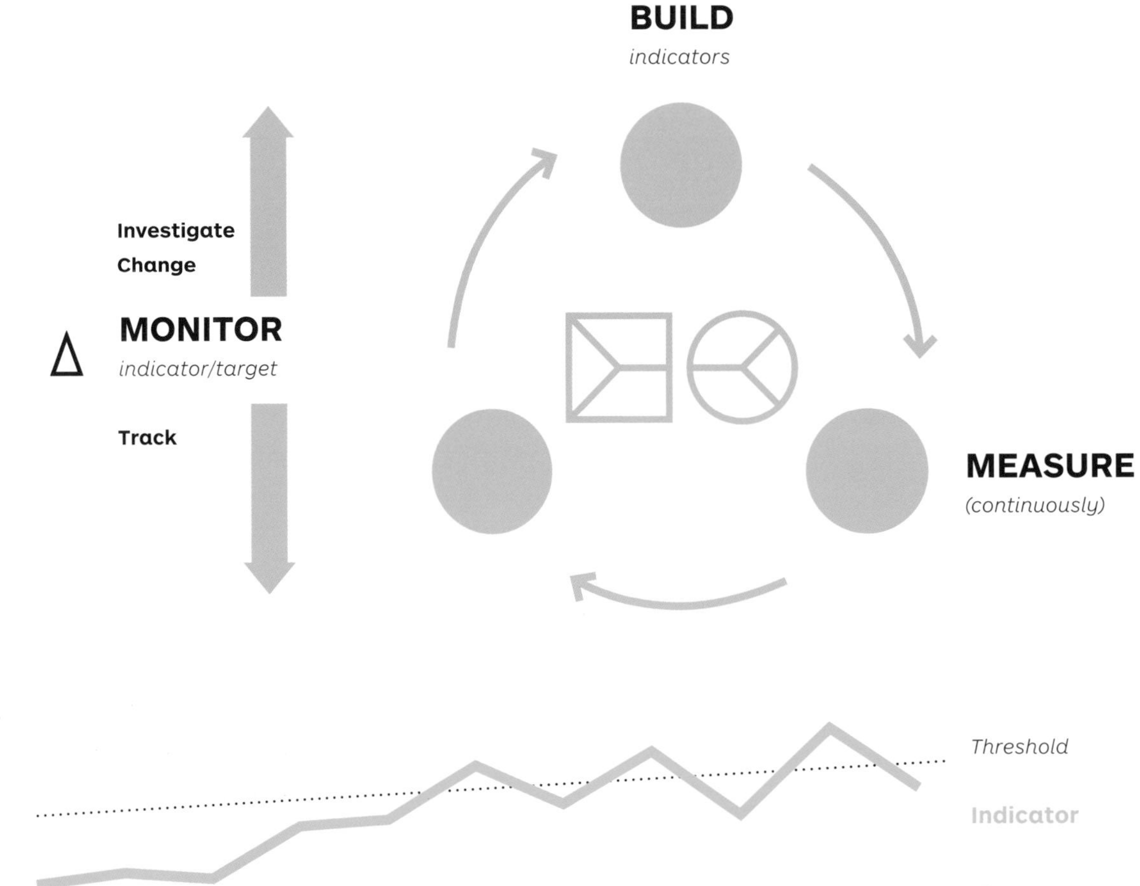

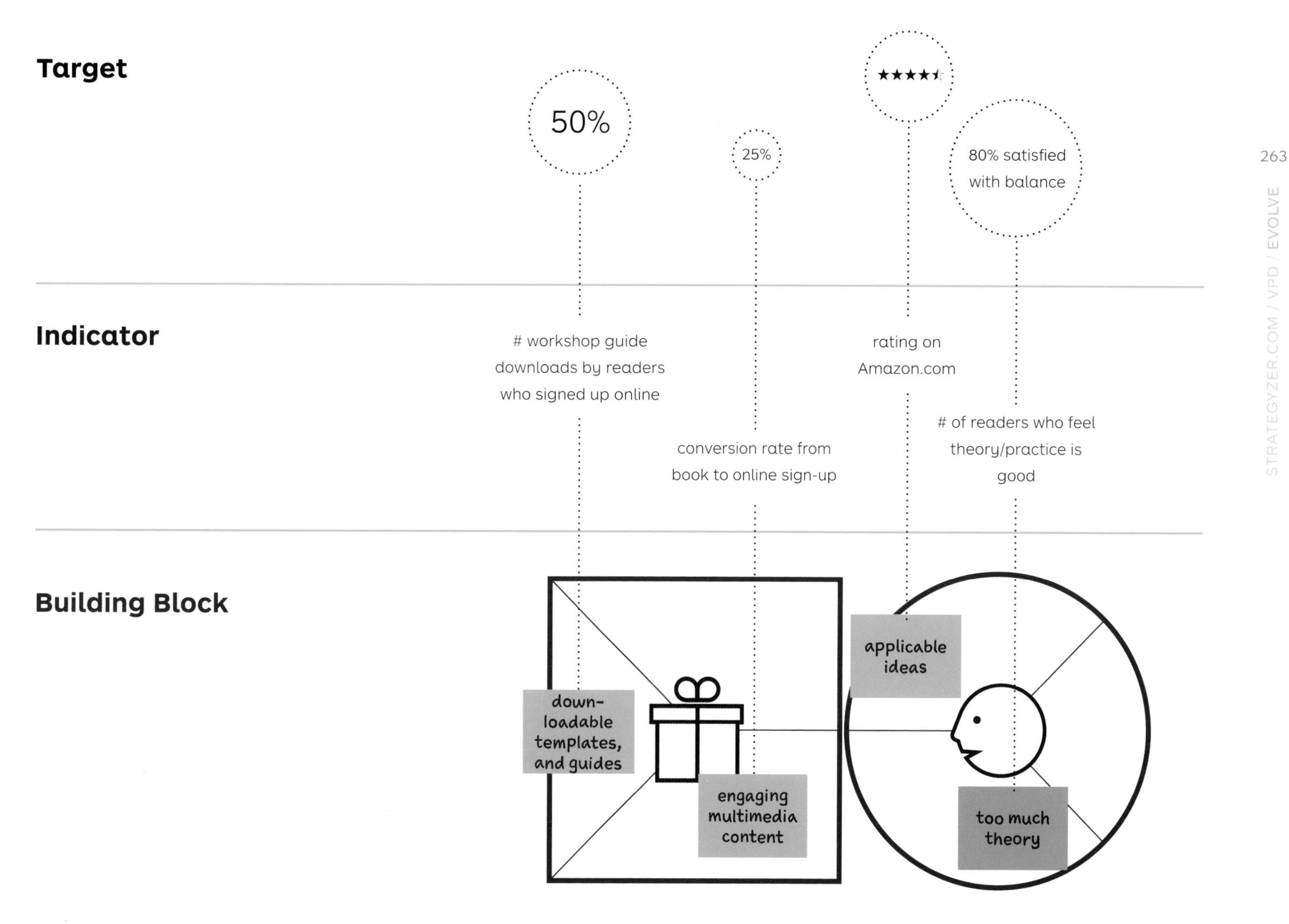
Target
50%
25%
★★★★☆
80% satisfied with balance
Indicator
workshop guide downloads by readers who signed up online
conversion rate from book to online sign-up
rating on Amazon.com
# of readers who feel theory/practice is good
Building Block
down-loadable templates, and guides
engaging multimedia content
applicable ideas
too much theory

Improve Relentlessly

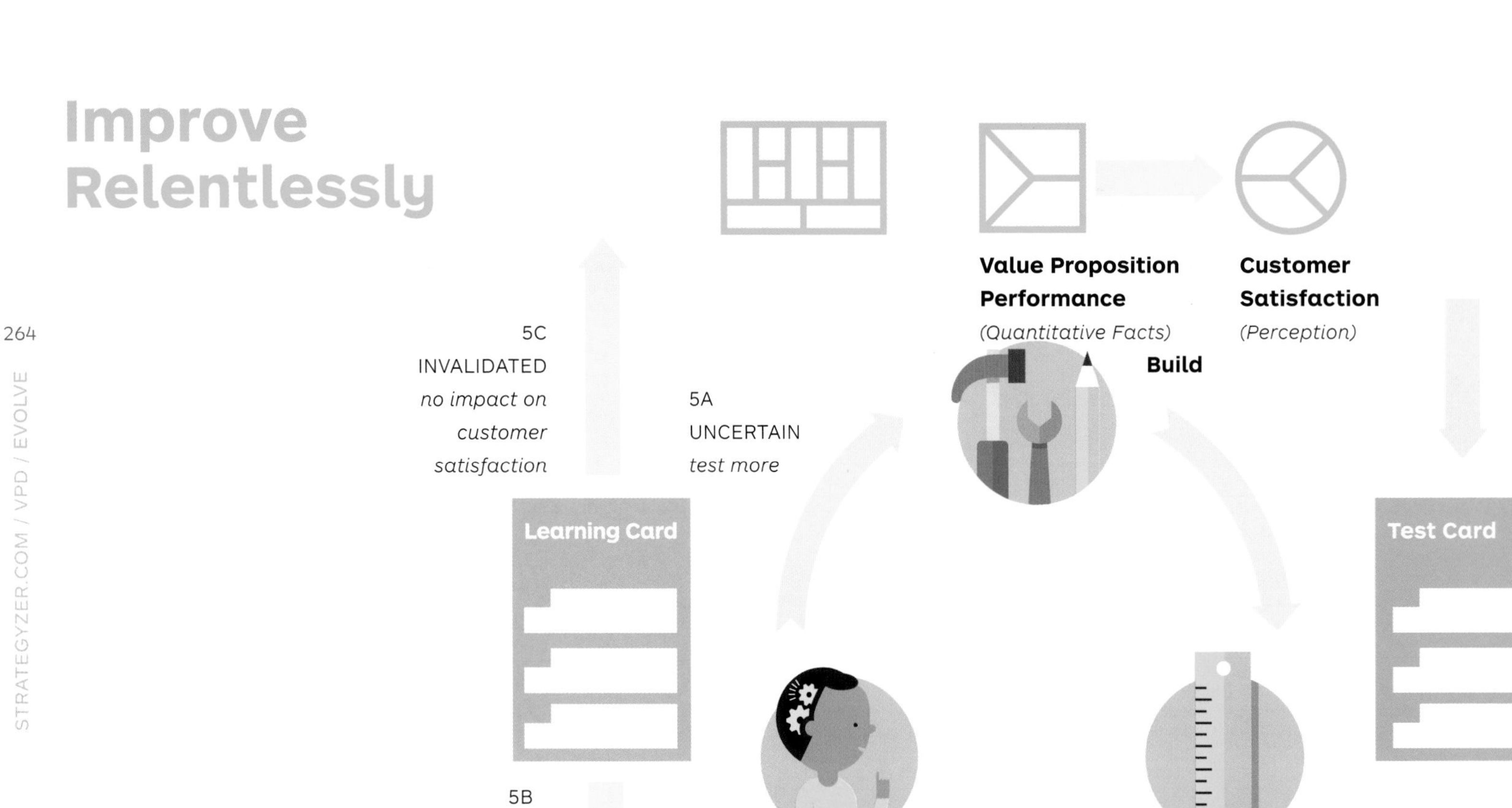

Use the same tools and processes from testing and monitoring to improve your value proposition once it's in the market. Continuously test "what if" improvement scenarios, and measure their impact on customer satisfaction.

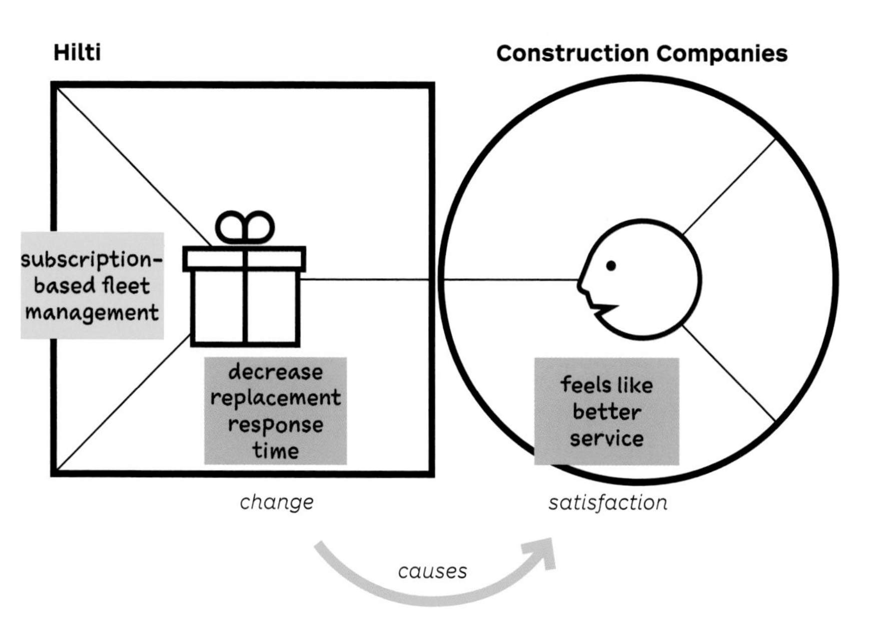

Test Card — Strategyzer

Test Name | Deadline
Assigned to | Duration

STEP 1: HYPOTHESIS
We believe that if we decrease the response time to replace broken tools, customers feel like they are getting a better service.
Critical:

STEP 2: TEST
To verify that, we will decrease response time for one client by 25% on average.
Test Cost: Data Reliability:

STEP 3: METRIC
And measure customer satisfaction at the beginning and the end of the experiment.
Time Required:

STEP 4: CRITERIA
We are right if customer satisfaction increases by x%.

Copyright Business Model Foundry AG — The makers of Business Model Generation and Strategyzer

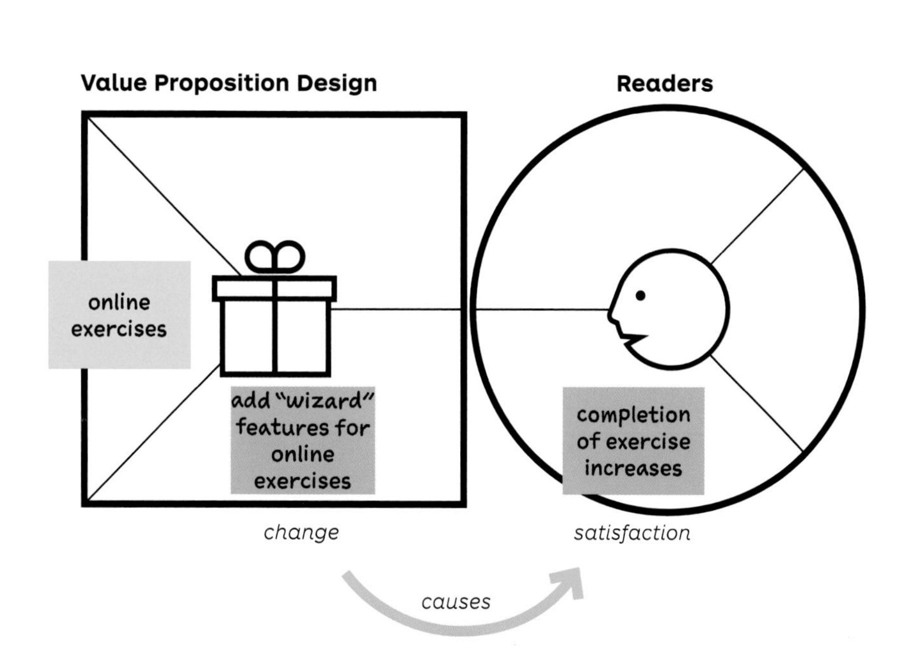

Reinvent Yourself Constantly

Successful companies create value propositions that sell embedded in business models that work. Outstanding companies do so continuously. They create new value propositions and business models while they are successful.

Today's enterprise must be agile and develop what Columbia Business School Professor Rita McGrath calls transient advantages in her book *The End of Competitive Advantage*. She argues that companies must develop the ability to rapidly and continuously address new opportunities, rather than search for increasingly unsustainable long-term competitive advantages.

Use the tools and processes of *Value Proposition Design* to continuously reinvent yourself and create new value propositions embedded in great business models.

Five things to remember when you build transient advantages:

- Take the exploration of new value propositions and business models just as seriously as the execution of existing ones.
- Invest in continuously experimenting with new value propositions and business models rather than making big bold uncertain bets.
- Reinvent yourself while you are successful; don't wait for a crisis to force you to.
- See new ideas and opportunities as a means to energize and mobilize employees and customers rather than a risky endeavor.
- Use customer experiments as a yardstick to judge new ideas and opportunities rather than the opinions of managers, strategists, or experts.

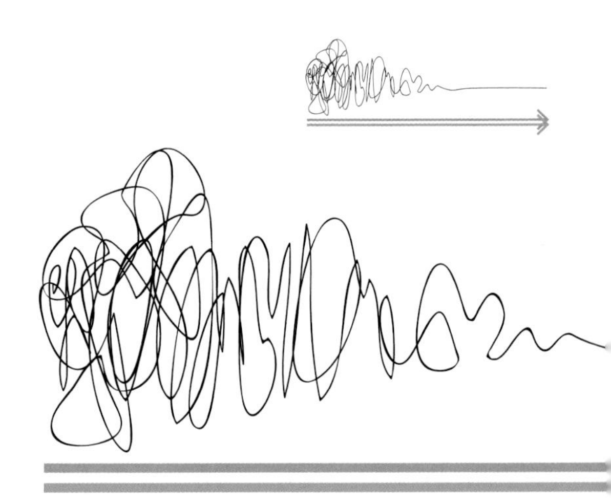

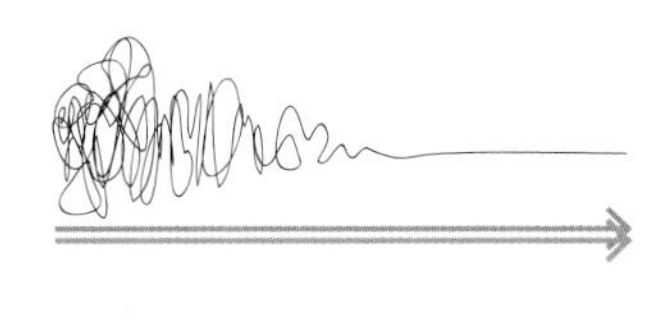

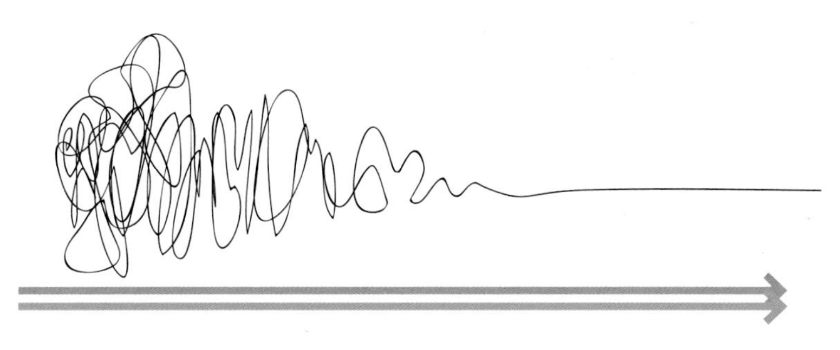

Continuously ask yourself...

What elements in your environment are changing? What do market, technology, regulatory, macroeconomic, or competitive changes mean for your value propositions and business models? Do those changes offer an opportunity to explore new possibilities or could they be a threat that might disrupt you?

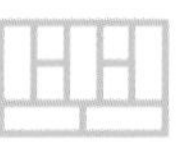

Is your business model expiring? Do you need to add new resources or activities? Do the existing ones offer an opportunity to expand your business model? Could you bolster your existing business model or should you build completely new ones? Is your business model portfolio fit for the future?

Today's enterprise must be agile and develop what Columbia Business School Professor Rita McGrath calls transient advantages in her book *The End of Competitive Advantage*. She argues that companies must develop the ability to rapidly and continuously address new opportunities, rather than search for increasingly unsustainable long-term competitive advantages.

Taobao: Reinventing (E-)Commerce

Taobao is the Chinese e-commerce phenomenon, part of the Alibaba Group. It is credited with ushering in a new wave of commerce in China by using the Internet to create an ecosystem where trusted commercial exchanges could take place. In 10 years it evolved its business models three times. It proactively embraced the changes taking place on its platform and in the wider Chinese economy and turned them into an opportunity.

Check out the full Taobao case online

2003

A new Consumer-to-Consumer (C2C) Platform

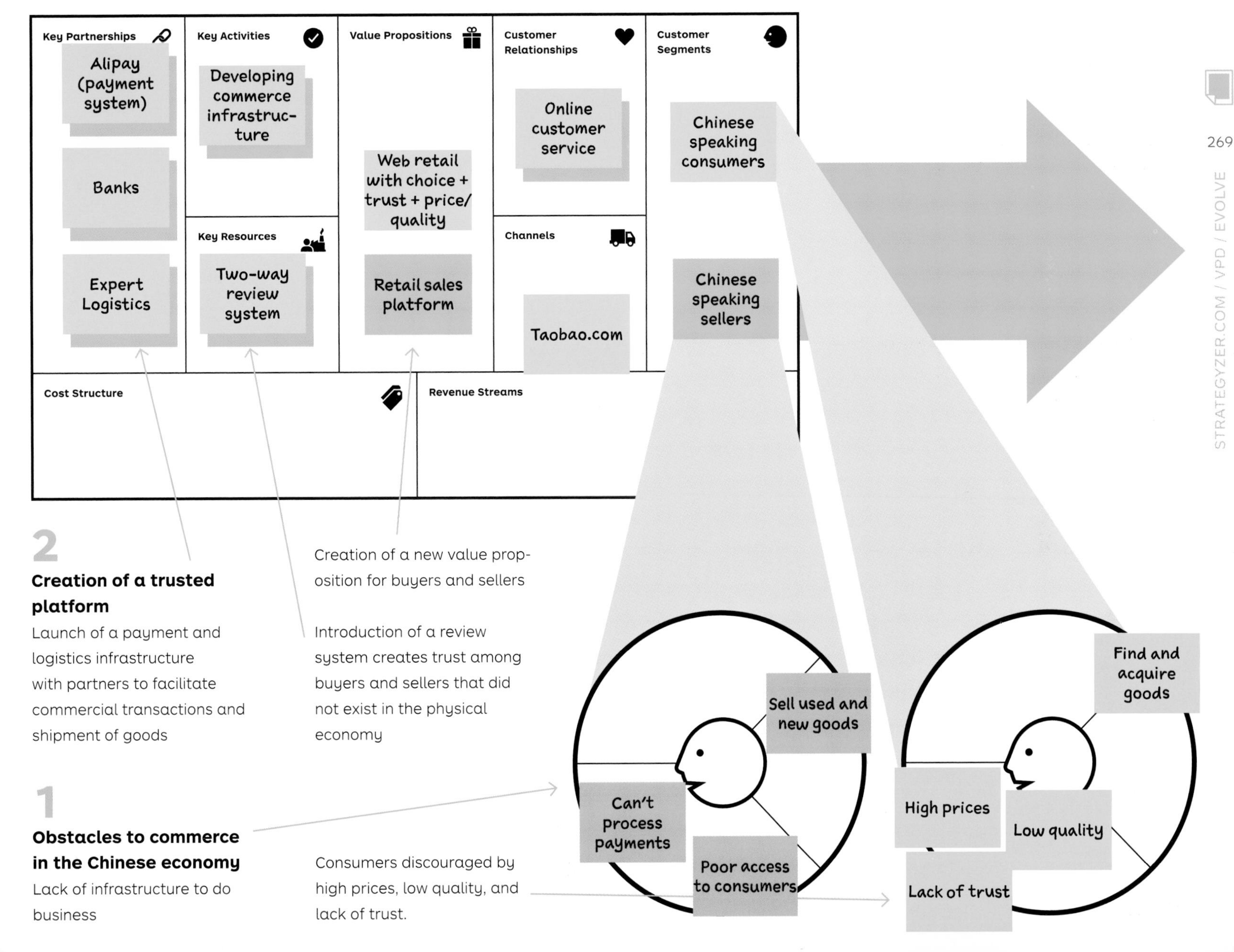

2006

Taobao — Small Business-to-Consumer (B2C)

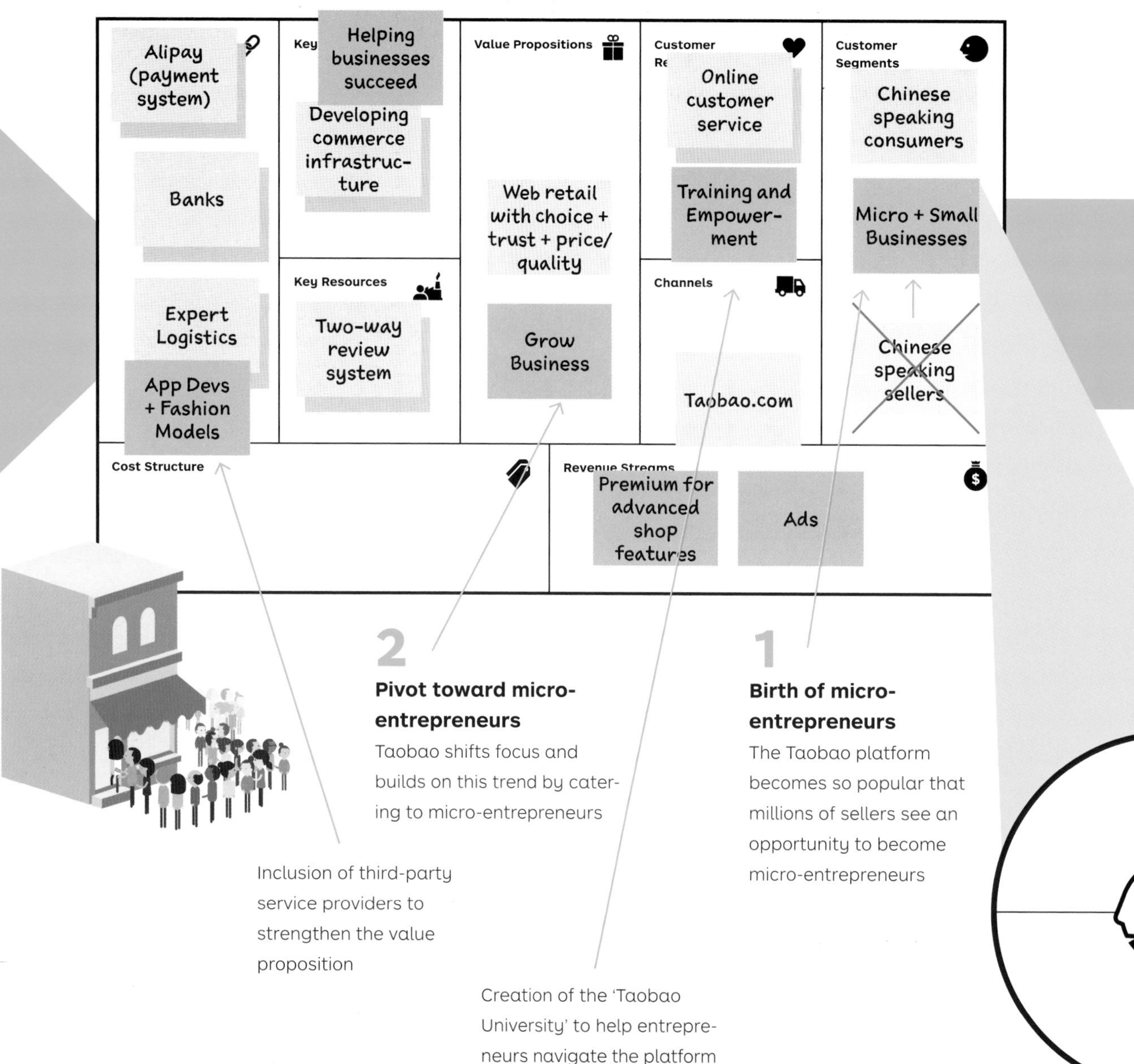

2

Pivot toward micro-entrepreneurs
Taobao shifts focus and builds on this trend by catering to micro-entrepreneurs

1

Birth of micro-entrepreneurs
The Taobao platform becomes so popular that millions of sellers see an opportunity to become micro-entrepreneurs

Inclusion of third-party service providers to strengthen the value proposition

Creation of the 'Taobao University' to help entrepreneurs navigate the platform and learn about business

2008
Taobao — Big Business-to-Consumer (B2C)

2013
Taobao — ?

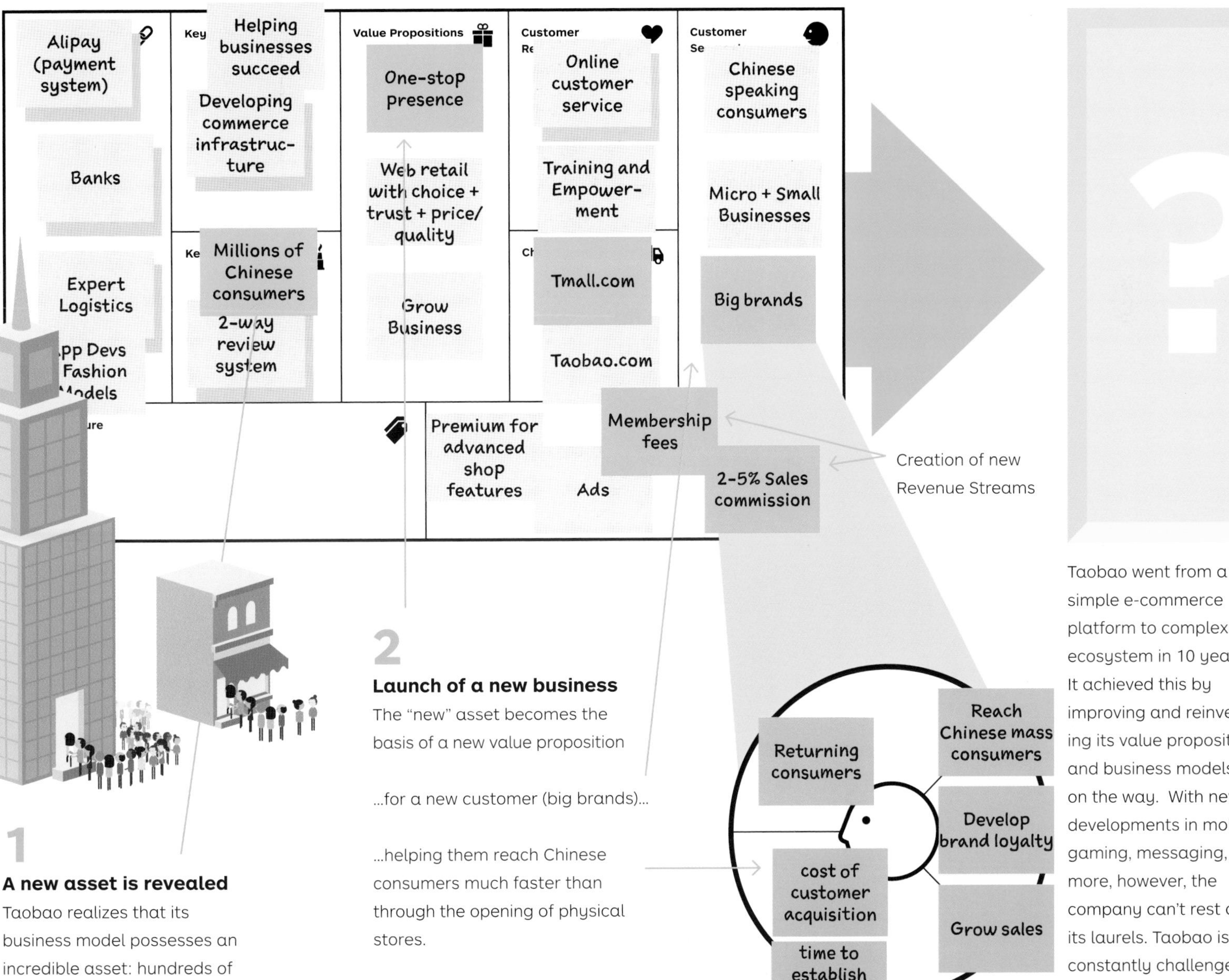

1

A new asset is revealed

Taobao realizes that its business model possesses an incredible asset: hundreds of millions of Chinese consumers.

2

Launch of a new business

The "new" asset becomes the basis of a new value proposition

...for a new customer (big brands)...

...helping them reach Chinese consumers much faster than through the opening of physical stores.

Taobao went from a simple e-commerce platform to complex ecosystem in 10 years. It achieved this by improving and reinventing its value propositions and business models on the way. With new developments in mobile, gaming, messaging, and more, however, the company can't rest on its laurels. Taobao is constantly challenged to continue its evolution.

Lessons Learned

Create Alignment

The Value Proposition and Business Model Canvases are excellent alignment tools. Use them as a shared language to create better collaboration across the different parts of your organization. Help every stakeholder understand how exactly you intend to create value for your customers and your business.

Measure, Monitor, Improve

Track the performance of your value propositions over time to make sure you continue to create customer value while market conditions change. Use the same tools and processes to improve your value propositions, which you used to design them.

Reinvent While Successful

Don't wait with reinventing your value propositions and business models. Do so before before market conditions force you to, because it might be too late. Create organizational structures that allow you to improve existing value propositions and business models and invent new ones at the same time.

after

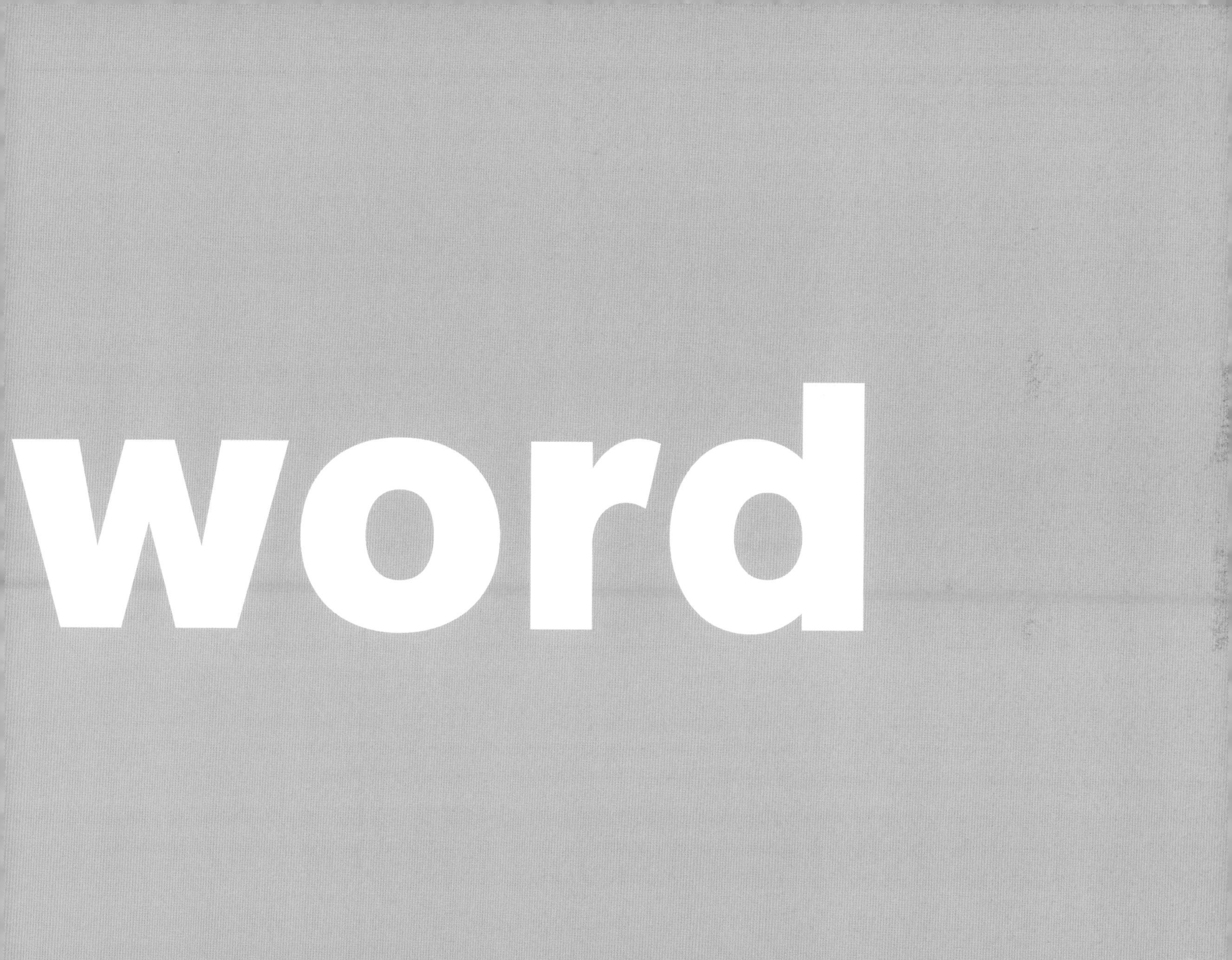
word

Glossary

(Business) Hypothesis
Something that needs to be true for your idea to work partially or fully but that hasn't been validated yet.

Business Model
Rationale of how an organization creates, delivers, and captures value.

Business Model Canvas
Strategic management tool to design, test, build, and manage (profitable and scalable) business models.

Call to Action (CTA)
Prompts a subject to perform an action; used in an experiment in order to test one or more hypotheses.

Customer Development
Four-step process invented by Steve Blank to reduce the risk and uncertainty in entrepreneurship by continuously testing the hypotheses underlying a business model with customers and stakeholders.

Customer Gains
Outcomes and benefits customers must have, expect, desire, or dream to achieve.

Customer Insight
Minor or major breakthrough in your customer understanding helping you design better value propositions and business models.

Customer Pains
Bad outcomes, risks, and obstacles that customers want to avoid, notably because they prevent them from getting a job done (well).

Customer Profile
Business tool that constitutes the right-hand side of the Value Proposition Canvas. Visualizes the jobs, pains, and gains of a customer segment (or stakeholder) you intend to create value for.

Environment Map
Strategic foresight tool to map the context in which you design and manage value propositions and business models.

Evidence
Proves or disproves a (business) hypothesis, customer insight, or belief about a value proposition, business model, or the environment.

Experiment/Test
A procedure to validate or invalidate a value proposition or business model hypothesis that produces evidence.

Fit
When the elements of your value map meet relevant jobs, pains, and gains of your customer segment and a substantial number of customers "hire" your value proposition to satisfy those jobs, pains, and gains.

Gain Creators
Describes how products and services create gains and help customers achieve the outcomes and benefits they require, expect, desire, or dream of by getting a job done (well).

Jobs to Be Done
What customers need, want, or desire to get done in their work and in their lives.

Lean Start-up
Approach by Eric Ries based on the Customer Development process to eliminate waste and uncertainty from product development by continuously building, testing, and learning in an iterative fashion.

Learning Card
Strategic learning tool to capture insights from research and experiments.

Minimum Viable Product (MVP)
A model of a value proposition designed specifically to validate or invalidate one or more hypotheses.

Pain Relievers
Describes how products and services alleviate customer pains by eliminating or reducing bad outcomes, risks, and obstacles that prevent customers from getting a job done (well).

Products and Services
The items that your value proposition is based on that your customers can see in your shop window—metaphorically speaking.

Progress Board
Strategic management tool to manage and monitor the business model and value proposition design process and track progress toward a successful value proposition and business model.

Prototyping (low/high fidelity)
The practice of building quick, inexpensive, and rough study models to learn about the desirability, feasibility, and viability of alternative value propositions and business models.

Test Card
Strategic testing tool to design and structure your research and experiments.

Value Map
Business tool that constitutes the left-hand side of the Value Proposition Canvas. Makes explicit how your products and services create value by alleviating pains and creating gains.

Value Proposition
Describes the benefits customers can expect from your products and services.

Value Proposition Canvas
Strategic management tool to design, test, build, and manage products and services. Fully integrates with the Business Model Canvas.

Value Proposition Design
The process of designing, testing, building, and managing value propositions over their entire lifecycle.

Core Team

Yves Pigneur
Supervising Author

Trish Papadakos
Designer

Greg Bernarda
Author

Alex Osterwalder
Lead Author
Strategyzer Cofounder

Alan Smith
Author + Creative Director
Strategyzer Cofounder

Tegan Mierle **Sarah Kim**

Brandon Ainsley **Matt Mancuso**

Pilot Interactive
Illustration Team

Strategyzer Content Team
Benson Garner, Nabila Amarsy

Strategyzer Product Team
Dave Lougheed, Tom Phillip, Joannou Ng, Chris Hopkins,
Matt Bullock, Federico Galindo

Prereaders

We practice what we preach and tested our ideas before releasing them. More than 100 selected people from around the world participated as prereaders to scrutinize our raw creations. More than 60 actively contributed by reviewing ideas, concepts, and spreads. They offered suggestions, meticulously proofread, and pointed out flaws and inconsistencies without pity. We iterated the book title several times with prereaders before testing various alternatives in the market.

Gabrielle Benefield
Phil Blake
Jasper Bouwsma
Frederic Briguet
Karl Burrow
Manuel Jose Carvajal
Pål Dahl
Christian Doll
Joseph Dougherty
Todd Dunn
Reinhard Ematinger
Sven Gakstatter
Jonas Giannini
Claus Gladyszak
Boris Golob
Dave Gray
Gaute Hagerup
Natasha Hanshaw
Chris Hill
Luke Hohmann
Jay Jayaraman
Shyam Jha
Greg Judelman
James King
Hans Kok
Ryuta Kono
Jens Korte
Jan Kyhnau
Michael Lachapelle
Ronna Lichtenberg
Justin Lokitz
Ranjan Malik
Deborah Mills-Scofield
Nathan Monk
Mario Morales
Fabio Nunes
Jan Ondrus
Aloys Osterwalder
Matty Paquay
Olivier Perez Kennedy
Johan Rapp
Christian Saclier
Andrea Schrick
Gregoire Serikoff
Aron Solomon
Peter Sonderegger
Lars Spicker Olesen
Matt Terrell
James Thomas
Paris Thomas
Patrick Van Der Pijl
Emanuela Vartolomei
Mauricio
Reiner Walter
Matt Wanat
Lu Wang
Marc Weber
Judith Wimmer
Shin Yamamoto

Bios

Alex Osterwalder

Dr. Alexander Osterwalder is the lead author of the international bestseller *Business Model Generation*, passionate entrepreneur, and demanded speaker. He cofounded Strategyzer, a software company specializing in tools and content for strategic management and innovation. Dr. Osterwalder invented the Business Model Canvas, the strategic management tool to design, test, build, and manage business models, which is used by companies like Coca Cola, GE, P&G, Mastercard, Ericsson, LEGO, and 3M. He is a frequent keynote speaker in leading organizations and top universities around the world, including Stanford, Berkeley, MIT, IESE, and IMD.
Follow him online @alexosterwalder.

Yves Pigneur

Dr. Yves Pigneur is coauthor of *Business Model Generation* and a professor of management and information systems at the University of Lausanne. He has held visiting professorships in the United States, Canada, and Singapore. Yves is a frequent guest speaker on business models in universities, large corporations, entrepreneurship events, and international conferences.

Greg Bernarda

Greg Bernarda is a thinker, creator, and facilitator who supports individuals, teams, and organizations with strategy and innovation. He works with inspired leaders to (re)design a future which employees, customers, and communities can recognize as their own. His projects have been with the likes of Colgate, Volkswagen, Harvard Business School, and Capgemini. Greg is a frequent speaker; he cofounded a series of events on sustainability in Beijing; and is an advisor at Utopies in Paris. Prior to that, he was at the World Economic Forum for eight years setting up initiatives for members to address global issues. He holds an MBA (Oxford Saïd) and is a Strategyzer certified business model coach.

Alan Smith

Alan is obsessed with design, business, and the ways we do them. A design-trained entrepreneur, he has worked across film, television, print, mobile, and web. Previously, he cofounded The Movement, an international design agency with offices in London, Toronto, and Geneva. He helped create the Value Proposition Canvas with Alex Osterwalder and Yves Pigneur, and the breakthrough design for *Business Model Generation*. He cofounded Strategyzer, where he builds tools and content with an amazing team; helping businesspeople make stuff customers want. Follow him online @thinksmith.

Trish Papadakos

Trish is a designer, photographer, and entrepreneur. She holds a Masters in Design from Central St. Martins in London and Bachelor of Design from the York Sheridan Joint Program in Toronto. She teaches design at her alma mater, has worked with award-winning agencies, launched several businesses, and is collaborating for the third time with the Strategyzer team. Follow her photography on Instagram @trishpapadakos.

Index

A

A/B testing, 230–231
ad-libs, for Prototyping, 76, 82–83
Ad Tracking (Experiment Library), 220
AirBnB, 91
Alibaba Group, 268–271
alignment, creating, 260–261
Anthropologist, 106, 114–115, 217
Apple, 156, 157
App Store (Apple), 157
assessment
  Business Model and, 156–157
  of competitors, 128–129, 130–131
  of skills for Value Proposition Design, xxii–xxiii
  of Value Proposition, 122–123
Azuri, 146–151

B

best practices
  for mapping customers' jobs, pains, and gains, 24–25 (*See also* Customer Gains; Customer Pains; Jobs to be Done)
  for mapping value creation, 30
Blank, Steve, 118, 182–183
Blue Ocean Strategy, 130
books, as Starting Points, 92–93
brainstorming
  defining criteria with, 140
  possibilities for, 92–93
  See also Starting Points
Bransfield-Garth, Simon, 146
brochures, creating, 222
Build, Measure, Learn Circle, 94, 95
Business Hypothesis
  defined, 201
  extracting, 200–201
  Lean Startup and, 185, 186–187
  prioritizing, 202–203
Business Model, 142–157
  assessing, 156–157
  Azuri example, 146–151
  compressed air energy storage example, 152–153
  creating value for customers and, 144–145
  Fit and, 48–49, 52–53
  platform Business Models, 52–53
  stress testing, 154–155
  testing, 194–195
Business Model Canvas
  defined, xv
  illustrated, xvii
Business Model Generation (Osterwalder), xiv, xvi
business plans, experimentation processes *versus*, 179
business-to-business (B2B) transactions, 50–51
Buy a Feature (Experiment Library), 235
Buyer of Value (customer role), 12

C

Call to Action (CTA), 218–219
Canvas
    Customer Profile and, 3–5, 9, 10–25
    Fit and, 3–5, 40–59
    identifying stakeholders with, 50–51
    moviegoing example, 54–55
    observing customers, 7
    summarized, 60
    Value Map and, 3–5, 8, 26–39
    Value Proposition, defined, 6
    See also Business Model Canvas; Customer Profile; Fit; Value Map; Value Proposition; Value Proposition Canvas
change. *See* Evolve
channels, defined, xvi
characteristics, of customers, 14–15
Choices, 120–141
    competitors and, 128–129, 130–131
    context and, 126–127
    defining criteria and selecting prototypes, 140–141
    feedback and, 132–133, 134–135, 136–137
    role-playing and, 107, 124–125
    tips for, 124, 131, 137
    Value Proposition assessment, 122–123
    visualization and, 138–139
Circle, Testing the, 190–191
Cocreator (for Customer Insight), 107
Cocreator of Value (customer role), 12
colleagues, Value Proposition Design for, xxiv–xxv
Company Building (Customer Development process), 183
competitors, assessing, 128–129, 130–131
compressed air energy storage example
    Business Model, 152–153
    Prototyping, 96–97
context
    Jobs to be Done, 13
    understanding, 126–127
cost structure, defined, xvi
Creation (Customer Development process), 183
criteria, defining, 140–141
Customer Development process, 182–183
Customer Gains
    best practices for mapping, 24–25
    checking Fit and, 46–47
    defined, 16–17
    psychodemographic profile approach *versus*, 54–55
    ranking, 20–21
    as Starting Point, 88–89
    Testing the Circle objective, 190–191
    See also Gain Creators
Customer Insight, 104–119
    Anthropologist, 106, 114–115, 217
    choosing mix of experiments for, 216–217 (*See also* Experiment Library)
    Cocreator, 107
    creating value for, 144–145 (*See also* Business Model)
    customer relationship management (CRM), xvi, 109
    Data Detective, 106, 108–109, 217

gaining, 106–107
identifying patterns in, 111, 116–119
Impersonator, 107, 124–125
Journalist, 106, 110–113, 217, 225
Scientist, 107
shaping ideas and, 70–71
tips for, 113, 115, 117
Customer Pains
best practices for mapping, 24–25
checking Fit and, 46–47
defined, 14–15
psychodemographic profile approach *versus*, 54–55
ranking, 20–21
as Starting Point, 88–89
Testing the Circle objective, 190–191
See also Pain Relievers
Customer Profile, 10–25
best practices for mapping jobs, pains, and gains, 24–25
business-to-business (B2B) transactions, 50–51
Customer Gains, defined, 16–17
Customer Pains, defined, 14–15
customer's context and, 56–57
customer segments, xvi, 116
defined, 9
different solutions for same customers, 58–59
identifying high-value jobs, 100–101
innovating from, 102–103
jobs, pains, and gains as new approach, 54–55
Jobs to be Done, defined, 12–13
ranking jobs, pains, and gains, 20–21
sketching, 18–19
understanding customer perspective for, 22–23
use of, 60–61
See also Jobs to be Done; Starting Points

D

Data Detective, 106, 108–109, 217
data mining, 109
data sheets, creating, 222
data traps, avoiding, 210–211
Day in the Life, A (worksheet), 115–116
de Bono, Edward, 136–137
decision makers, 50–51
Dell, 157
Design, 64–170
characteristics of great Value Propositions, 72–73
constraints, 90–91
design/build (Lean Startup), 185, 186–187
in established organizations, 158–187 (*See also* established organizations)
finding the right Business Model, 142–157 (*See also* business model)
making choices for, 120–141 (*See also* Choices)
overview, 67
Prototyping possibilities, 70–71, 74–85 (*See also* Prototyping)
shaping ideas with, 70–71
Starting Points, 70–71, 86–103 (*See also* Starting Points)
summarized, 170
understanding customers, 70–71, 104–119 (*See also* Customer Insight)
discovery (Customer Development process), 182
Dotmocracy, 138–139
Dropbox, 210

E

Earlyvangelists, 118
economic buyers, 50–51
Eight19 (Azuri), 146–151
emotional jobs, of customers, 12
end users, 50–51. See also Customer Insight
Environment Map, v, xv
EPFL, 96–97
established organizations, 158–169
  inventing and improving, 160–161, 162–163
  reinventing, 164–165
  Value Proposition Design for, xix
  workshops for, 166–167, 168–169
Evidence
  Call to Action (CTA) and, 218–219
  need for, 190–195
  producing, 97, 216
  See also Test
Evolve, 254–272
  creating alignment and, 260–261
  improvement and, 264–265
  measuring and monitoring, 262–263
  overview, 257
  reinventing and, 266–267, 268–271
  summarized, 272
"exhausted maximum" trap, 211
experience, as feedback, 134
Experiment Library, 214–237
  Ad Tracking, 220
  Buy a Feature, 235
  Call to Action (CTA), 218–219
  choosing mix of experiments, 216–217
  experiment, defined, 216
  experiment design and, 204–205
  experimenting to reduce risk, 178–179
  Illustrations, Storyboards, and Scenarios, 222, 224–225
  Innovation Games, 232
  Landing Page MVP, 228–229
  Life-Size Experiments, 226–227
  Minimum Viable Product (MVP), 222–223, 228–229
  Mock Sales, 236–237
  Pre Sales, 237
  Product Box, 234
  Speed Boat, 233
  Split Testing, 230–231
  tips for, 217, 222, 224, 227, 229, 231, 233, 237
  Unique Link Tracking, 221
  See also Test

F

Facebook, 157
facts, as feedback, 134
false-negative/false-positive traps, 210
Federal Institute of Technology (Switzerland), 96–97
feedback, 132–133, 134–135, 136–137
financial issues
  cost structure, defined, xvi
  generating revenue, 144–145
  profit, defined, xvi

revenue streams, defined, xvi
stress testing, 154–155
testing customers' willingness to pay, 219
(*See also* Experiment Library)
See also Business Model
Fit, 40–59
addressing customers' jobs, pains, and gains with, 44–45
checking for, 46–47
Customer Profile and customer context, 56–57
Customer Profile and Value Map as two sides of, 3–5
Customer Profile *versus* psychodemographic profile approach, 54–55
different solutions for same customers, 58–59
multiple Fit, 52–53
stages of, 48–49
striving for, 42–43
use of, 60–61
functional jobs, of customers, 12

G

Gain Creators
Fit and, 9, 47
Pain Relievers *versus*, 38
Products and Services as, 33
Value Map and, 33–34
gains. *See* Customer Gains
Google
AdWords, 220
searches, 108
government census data, 108

H

Hilti, 90
Hohmann, Luke, 232
hypothesis. *See* Business Hypothesis

I

Ikea, 157
Illustrations, Storyboards, and Scenarios (Experiment Library), 222, 224–225
Impersonator, 107
improvement
for established organizations, 160–161, 162–163
as relentless, 264–265
See also Evolve
Indigo, 150–151
influencers, 50–51
Innovation Games, 232
intermediary Fit, 52–53
interviewing, of customers, 106, 110–113, 217, 225
iPod (Apple), 156

J

Jobs to be Done
best practices for mapping, 24–25
defined, 12–13
identifying high-value jobs, 98–99, 100–101
psychodemographic profile approach *versus*, 54–55

ranking, 20–21
as Starting Point, 88–89
Testing the Circle objective, 190–191
Journalist, 106, 110–113, 217, 225

K

key activities, defined, xvi
key partnerships, defined, xvi
key resources, defined, xvi

L

Landing Page MVP (Experiment Library), 228–229
Lean Startup, 184–185, 186–187
Learning Cards, 206–207, 213
learning (Lean Startup), 185, 186–187
Life-Size Experiments (Experiment Library), 226–227
listening, 112
Lit Motors, 226
"local maximum" trap, 211

M

magazines, as Starting Points, 92–93
market pull, 95
Marriott, 227
measurement
Evolve and, 262–263
Lean Startup feature, 185, 186–187
See also Test
MedTech, 154–155
Minimum Viable Product (MVP)
Lean Start-up with, 184
Prototyping and, 77
testing with, 222–223, 228–229
Mock Sales (Experiment Library), 236–237

N

napkin sketch, for Prototyping, 76, 80–81
Nespresso, 90, 156
new ventures, Value Proposition Design for, xviii

O

observation, of customers, 106, 114–115, 216–217. *See also* Experiment Library
obstacles, of customers, 14–15. *See also* Customer Pains
opinion, as feedback, 134
Osterwalder, Alexander, xiv, xvi
outcomes
Customer Gains as, 16–17
undesired, by customers, 14–15
Owlet, 246–251

P

Pain Relievers
  Fit and, 9, 47
  Gain Creators *versus*, 38
  Products and Services as, 31–32
  Value Map and, 33–34
"participatory tv," for understanding context, 126, 127
patterns, identifying, 111, 116–119
personal jobs, of customers, 12
perspective, of customers, 22–23. *See also* Customer Insight
platform Business Models, 52–53
Pre Sales (Experiment Library), 237
prioritization, 202–203, 219.
  See also Experiment Library
problems, of customers, 14–15.
  See also Customer Pains
Problem-Solution Fit, 48–49
Product Box (Experiment Library), 234
Product-Market Fit, 48–49
Products and Services
  meeting customer expectations with, 31–32
  multiple Fit for, 52–53
  Testing the Square objective, 192–193
  types of, 29–30
  value of, to customers, 31–32
profit, defined, xvi
Progress Board, 242–243, 244–245
Prototyping, 74–85
  ad-libs for, 76, 82–83
  defined, 76
  napkin sketch for, 76, 80–81
  packaging and, 223, 234
  principles of, 78–79
  selecting prototypes, 140–141
  shaping ideas with, 70–71
  spaces for, 227
  tips for, 77
  Value Proposition Canvas for, 77, 84–85
  See also Experiment Library
psychodemographic profiles, as traditional approach, 54–55
push *versus* pull debate, 94–95, 96–97, 98–99, 100–101

R

ranking, for customer jobs/pains/gains, 20–21
recommenders, 50–51
Rectangle, Testing the, 194–195
reinventing, to Evolve, 266–267, 268–271
required gains, of customers, 16
research, about customers. *See* Customer Insight
revenue streams, defined, xvi
Ries, Eric, 184–185
risks, customers and, 14–15
role-playing, 107, 124–125

S

saboteurs, 50–51
Scientist, 107

segmentation, xvi, 116
services. See Products and Services
Skype, 157
social jobs, of customers, 12
social media analytics, 109
Southwest, 91
Speed Boat (Experiment Library), 233
Split Testing (Experiment Library), 230–231
Square, Testing the, 192–193
stakeholders
  identifying, 50–51
  role-playing and, 107, 124–125
Starting Points, 86–103
  addressing customer concerns with, 88–89
  books and magazines for, 92–93
  design constraints and, 90–91
  innovation with Customer Profile, 102–103
  push *versus* pull debate, 94–95, 96–97, 98–99, 100–101
  shaping ideas with, 70–71
  tips for, 93, 97
start-ups, Value Proposition Design for, xviii
Step-by-Step Testing, 196–213
  avoiding data traps, 210–211
  extracting hypothesis, 200–201
  Learning Cards for insight, 206–207, 213
  learning speed and, 208–209
  overview, 198–199
  prioritizing hypothesis, 202–203
  Test Cards for experiment design, 204–205, 212
  tips for, 210
Strategy Canvas, 129, 130
Strategyzer logo, explained, x
stress testing, 154–155
supporting jobs, of customers, 12
Swatch, 90
synthesis, 116, 117

T

Taobao, 268–271
technology push, 94, 96–97
Test, 172–252
  Customer Development process, 182–183
  experimenting to reduce risk, 178–179
  Experiment Library for, 214–237
    (*See also* Experiment Library)
  Lean Startup, applying, 186–187
  Lean Startup movement, 184–185
  principles, 180–181
  Progress Board, 242–243, 244–245
  step-by-step, 196–213
    (*See also* Step-by-Step Testing)
  systematic design and testing example, 246–251
  testing process, summarized, 240–241, 252
  Testing the Circle objective, 190–191
  Testing the Rectangle objective, 194–195
  Testing the Square objective, 192–193
  tips for, 183
Test Cards, 204–205, 212

thinking hats (de Bono), 136–137
third-party research reports, 108
Transferrer of Value (customer role), 12

U

unique link tracking (Experiment Library), 221

V

validation (Customer Development process), 182
Value Map, 26–39
  best practices for mapping value creation, 30
  Gain Creators and, 33–34
  mapping how products and services create value, 36–38
  mapping value propositions, 34–35
  Pain Relievers, 31–32
  Products and Services, 29–30
  use of, 60–61
Value Proposition
  assessing competitors and, 128–129, 130–131
  assessment, 122–123
  Business Model connection to, 152–153
  defined, vi, xvi, 6
  See also Value Proposition Canvas; Value Proposition Design
Value Proposition Canvas
  characteristics of great value propositions, 72–73 (*See also* Design)
  for Prototyping, 77, 84–85
Value Proposition Design
  book organization and online companion, x
  Business Model Canvas, defined, xv (*See also* Business Model Canvas)
  competitors *versus*, 128–129
  Environment Map and, v, xv
  for established organizations, xix
  to overcome problems, vi–vii
  selling colleagues on, xxiv–xxv
  skills needed for, xxii–xxiii
  for start-ups, xviii
  successful use of, viii, xi
  tools and process of, xii–xiii (*See also* Canvas; Design; Evolve; Test)
  uses of, xx–xxi
  Value Proposition Canvas, defined, xiv, xv (*See also* Value Proposition Canvas)
visualization, 138–139

W

website
  tracking customers on, 109
  Value Proposition Design Online Companion, explained, x
websites
  landing pages of, 223, 228–229
WhatsApp, 157
workshops, for established organizations, 166–167, 168–169

Don't risk wasting your time, energy, and money working on products and services nobody wants— Flip the book over!